SYSTEMS BIOLOGY - THEORY, TECHNIQUES AND APPLICATIONS

FUNCTIONAL GENOMICS

NOVEL INSIGHTS, APPLICATIONS AND FUTURE CHALLENGES

HOLGER UFFE
AND
OLAF PHILIP
EDITORS

NOTICE TO THE READER

Additional color graphics may be available in the e-book version of this book.

Library of Congress Cataloging-in-Publication Data

ISBN: 978-1-53612-565-8

Published by Nova Science Publishers, Inc. † New York

SYSTEMS BIOLOGY - THEORY, TECHNIQUES AND APPLICATIONS

FUNCTIONAL GENOMICS

NOVEL INSIGHTS, APPLICATIONS AND FUTURE CHALLENGES

SYSTEMS BIOLOGY - THEORY, TECHNIQUES AND APPLICATIONS

Additional books in this series can be found on Nova's website under the Series tab.

Additional e-books in this series can be found on Nova's website under the eBooks tab.

CONTENTS

PREFACE

In Chapter One, the authors discuss the gene silencing tool RNA interference, or RNAi, which functions at both transcriptional and post-transcriptional levels. In Chapter Two, the authors study the origins of genome engineering, including its adaptation from prokaryotic to eukaryotic systems and applications. In Chapter Three, the authors discuss the utilization of different biotechnologies for insect pest management in plants, as well as give insight into recently developed approaches of genome editing for high throughput pest management. Closing the book with Chapter Four, the authors discuss how vital genetic resources, such as the ability to withstand scarcity of nutrients and water as well as to withstand pests and diseases, may be secured for future use in order to improve crops.

Chapter 1 – RNA interference (RNAi) is a gene silencing tool which functions at both transcriptional and post-transcriptional levels. Small non-coding RNAs of nearly 21-26 nucleotides are the actual players responsible for gene silencing. These small RNAs in a cell are derived from longer precursor molecules through a series of steps, each requiring a set of enzymatic and non enzymatic proteins. RNAi technology was first discovered in plants (Petunia) and subsequently in many metazoans (animals). The technology has turned out to be a promising tool in functional genomics, particularly for deciphering gene functions. It is also used in parallel with insecticidal protein coding genes (*cry*, *vip*, alpha amylase inhibitors, protease inhibitor proteins) for the development of insect tolerant crops, as sporadic emergence of pests is a major drawback in Bt crops. There are various successful examples in different crop species where *cry* genes from bacteria *Bacillus thuringiensis* have been extensively used for the development of insect tolerant crops against most of the lepidopteran and coleopteran pests. But cry genes suffer from a few limitations; till date, no *cry* gene is available for hemipteran insects and pests (having piercing and sucking type of mouth parts) and there are reports of development of insect resistance against cry genes. These limitations can be addressed by employing RNAi technology along with insecticidal genes for the development of insect-proof crop plants. This chapter will highlight the molecular mechanisms behind RNAi along with its applications and limitations in developing insect/pest tolerant crops.

Chapter 2 – Precise genome modification, in a predictable manner holds the key to bring in a new dawn in basic and applied research. Targeted genome editing using engineered nucleases is fast emerging as the foremost technique for elucidating and manipulating gene functions. Rapid advances in genome editing are being made after the emergence of the clustered, regularly interspaced, short palindromic repeats (CRISPR) – CRISPR-associated protein9 (Cas9) technology. This technique has been adapted from the microbial adaptive immune system, and is now being extensively used in developing different human therapeutics, alleviating genetic disorders, livestock breeding and developing improved high yielding plants. The system exploits the property of the RNA-guided endonuclease (Cas9) to cleave double stranded DNA in the targeted genomic loci under the direction of easily reprogrammable guide RNAs (gRNAs). Once the nicks are induced, they are either repaired by the cells' nonhomologous end-joining (NHEJ) repair mechanism, or sequence replacement by homology-directed repair (HDR). This technology can also be used to stimulate the function of a gene by fusing its activation domain to a catalytically inactive Cas9 and targeting the promoter region. This chapter shall in detail cover the origin of this technology, its adaptation from prokaryotic to eukaryotic systems, evolution and applications. The use of this system for functional genomics studies and applications of CRISPR-Cas9 in crop improvement will also be described. CRISPR-Cas9 has evolved to become the face of genetic engineering, and is set to revolutionize functional genomics in the near future.

Chapter 3 – Modern agriculture is already facing challenges of feeding the increasing global population, and many more difficulties lie ahead in the future. Mitigation of insects and pests is a herculean task so as to sustain productivity. This requires better pest management strategies, which would have a huge impact on the final productivity. This chapter discusses pest management strategies practiced over the past, such as agronomical approaches, modern breeding approaches, and use of chemical pesticides; for better defense strategies against key insect pests responsible for large loss of economics in terms of crop production. Various approaches based on genetic engineering including transgenics, and RNAi for insect resistance in plants have been described and followed. Utility of different biotechnological tools in the field of insect pest management will be discussed in this chapter, considering both benefits as well as limitations. Additionally, this chapter also puts forward an insight into recently developed novel approaches of genome editing for high throughput pest management.

Chapter 4 – Wild relatives of crops (WRC) are the ancestors of crop plants and other species closely related to crops, which are seen to be continuously growing under harsh conditions. Wild relatives of plants are differentially exposed to nature and live by different rules of adaptability and survival than their cultivated counterparts. Continuous exposure to severe climatic conditions has helped them evolve at the genetic level, making them more genetically diverse. These wild species are considered as one of the most important sources of resistance genes to tackle both biotic and abiotic stresses. Due to their broader genetic base, wild species have a potential role in the development of crops which can withstand adverse impacts of climate change, such as increasing scarcity of nutrients, water and other inputs, and increased incidents of new pests and diseases. These vital genetic resources can be secured for future use. Various morphological, physiochemical, biochemical and molecular mechanisms are responsible for making wild crop relatives resistant to various stresses. In this chapter the important biotic stress responsive traits possessed by wild relatives, specifically against insect pests, and their potential uses in crop improvement programmes are discussed.

In: Functional Genomics ISBN: 978-1-53612-565-8
Editors: Holger Uffe and Olaf Philip

Chapter 1

RNAi Technology to Combat Insect Pests in Plants

Ravi Prakash Saini[1], Kesiraju Karthik[1], Suraj Singh[1], N. Muralimohan[1], T. Arulprakash[1], Nikhil Ramkumar[1], Shweta Singh[1], Pragya Mishra[1], R. Maniraj[1], Saurabh Tyagi[1], Era Vaidya Malhotra[1], T. Vinutha[2], Debasis Pattanayak[1] and Rohini Sreevathsa[1,*]

[1]ICAR-National Research Centre on Plant Biotechnology, New Delhi, India

[2]Division of Biochemistry, ICAR-Indian Agricultural Research Institute, New Delhi, India

Abstract

RNA interference (RNAi) is a gene silencing tool which functions at both transcriptional and post-transcriptional levels. Small non-coding RNAs of nearly 21-26 nucleotides are the actual players responsible for gene silencing. These small RNAs in a cell are derived from longer

* Corresponding author Email: rohinisreevathsa@gmail.com.

precursor molecules through a series of steps, each requiring a set of enzymatic and non enzymatic proteins. RNAi technology was first discovered in plants (Petunia) and subsequently in many metazoans (animals). The technology has turned out to be a promising tool in functional genomics, particularly for deciphering gene functions. It is also used in parallel with insecticidal protein coding genes (*cry*, *vip*, alpha amylase inhibitors, protease inhibitor proteins) for the development of insect tolerant crops, as sporadic emergence of pests is a major drawback in Bt crops. There are various successful examples in different crop species where *cry* genes from bacteria *Bacillus thuringiensis* have been extensively used for the development of insect tolerant crops against most of the lepidopteran and coleopteran pests. But cry genes suffer from a few limitations; till date, no *cry* gene is available for hemipteran insects and pests (having piercing and sucking type of mouth parts) and there are reports of development of insect resistance against cry genes. These limitations can be addressed by employing RNAi technology along with insecticidal genes for the development of insect-proof crop plants. This chapter will highlight the molecular mechanisms behind RNAi along with its applications and limitations in developing insect/pest tolerant crops.

Keywords: RNAi, dsRNA, RISC complex, HD-RNAi, artificial microRNA

1. INTRODUCTION

Agriculture is an industry that humans heavily rely on for not only sustenance, but also other resources like clothing and shelter. Agriculture is the only industry that can provide food, which is essential to feed the growing world population. The availability of productive agricultural fields and water resources is getting narrower day by day due to various human practices. Thus, there is an urgent need for the development of new agricultural techniques for the maintenance of sustainable agricultural practises to bridge the yield gap and to increase the net productivity of food grains throughout the world. Agricultural productivity is affected by several biotic and abiotic stresses. Among biotic stresses, insect pests tend to be the main cause for loss of agricultural productivity. Chemical pesticides have been most extensively used to control insect pests for decades. With the advancement of knowledge, the adverse effects of chemical pesticides in causing pollution of ecological habitats including land and water resources have been well emphasised. Insect pests have acquired much resistance in the recent past as compared to that acquired over the past several decades due to the increased and uncontrolled use of chemical pesticides over and above the recommended doses.

Genetic engineering is a proven powerful alternative to overcome problems due to the use of chemical pesticides and involves development of plants that have an acquired resistance to overcome the insect pest attack. Transgenic plants harboring *Bacillus thuringiensis* (Bt) cry protein genes have emerged as a promising strategy and several Bt crops like cotton, soya bean, tomato, brinjal, cabbage, cauliflower etc., have been developed. Once insects feed upon these crops, rupturing of their intestinal epithelial cells occurs due to an interaction with the cry protein and its receptor. In spite of the specificity between larval receptors and cry proteins, some insect species have developed resistance to the *cry* genes and outbreak of secondary pests seems to be a major and direct impact of Bt technology throughout the world, thus emphasising on the importance of new biotechnological approaches to control these insect pests.

Plant biotechnologists around the world have been continuously striving to improve the major food crops by using various biotechnological applications. Some of them include - trait identification strategies, phenotypic studies, biotic and abiotic stress tolerance studies which would eventually lead to the development of genetically modified or transgenic plants carrying the desired trait of interest. RNA interference has emerged as a potential tool to mitigate insect menace in plants. The strategy uses the silencing of specific genes instigated by the presence of double stranded RNA molecules in the insects upon ingestion.

2. Definition of RNA Interference

RNA interference (RNAi) is a sequence specific gene silencing phenomenon mediated by small dsRNA molecules. It is a biological process in which small RNA molecules act as elements inhibiting gene expression or translation by the cleavage of targeted mRNA molecules. It is also known as co-suppression, post-transcriptional gene silencing (PTGS), and quelling. This natural mechanism of sequence-specific gene silencing is found in both plants and animals and can be exploited as an alternative approach to produce transgenic plants for insect resistance. This technique holds potential to revolutionize functional genomics, phenotyping, agriculture, therapeutic interventions, and other areas of research.

The discovery of RNAi was preceeded by a number of related observations. The first in line were the observations of transcriptional inhibition of antisense RNA expressed in transgenic plants (Davis et al., 1986) followed by reports of unexpected outcomes of experiments performed by plant scientists in the United States and Netherlands in 1990s (Napoli et al., 1990). Researchers in an attempt to alter flower colours in petunias introduced additional copies of the gene encoding chalcone synthase, an enzyme responsible for flower pigmentation, into petunia plants having normal pink or violet coloured flowers. The over expressed genes were expected to produce darker coloured flowers in the transgenic plants. But unexpectedly they produced less pigmented, fully or partially white flowers, which indicated that there was a reduction in the activity of chalcone synthase in the transgenic plants; in fact, both the endogenous genes and the transgenes were found to be down-regulated in the white flowers. This phenomenon was called co-suppression of gene expression (Romano et al., 1992).

A related event, termed quelling was observed in the fungus *Neurospora crassa.* Further investigation of this phenomenon in plants indicated that the down regulation was due to post-transcriptional inhibition of gene expression by an increased rate of mRNA degradation. Plant virologists working to improve plant resistance to viral diseases observed a phenomenon in which plants expressing virus-specific proteins showed enhanced tolerance to viral infection. However, it was not known that plants carrying some non-coding regions of viral RNA sequences would show similar levels of protection. Researchers believed that viral

RNA produced by transgenes could inhibit viral replication (Covey et al., 1997). In the reverse experiments, short sequences of plant genes were integrated into viruses, resulting in the suppression of targeted gene in the infected plant. This phenomenon was known as "virus-induced gene silencing" (VIGS), and such phenomena were called post-transcriptional gene silencing (Ratcliff et al., 1997). Taking cue from the initial reports in plants, researchers started looking for the presence of a similar phenomenon in other organisms (Guo et al., 1995, Pal-Bhadra et al., 1997).

In 1998, American scientists Andrew Z. Fire and Craig C. Mello discovered gene silencing by interfering RNA in *Caenorhabditis elegans*, a nematode, and shared the Nobel Prize (2006) in Physiology or Medicine for their contribution. They were able to successfully inhibit the transcriptional expression of specific genes by introducing short double-stranded RNA (dsRNA) segments into the cells of *Caenorhabditis elegans*. These dsRNA segments underwent enzymatic processing producing smaller double-stranded RNA that attached to RNA complexes and guided them towards the target messenger RNA (mRNA) that possessed the complementary nucleotide sequences. The attachment of both RNAs inhibited the translation of mRNA molecules into proteins or resulted in their cleavage. Their investigation reported that double-stranded RNA successfully silenced the targeted gene in the tissue, and thus the term RNAi was coined.

2.1. RNAi in Different Organisms

Since the discovery of RNAi and its regulatory potentials, it has become evident that RNAi has immense potential to suppress desired genes. The RNAi pathway is commonly found in many eukaryotes as a natural process that takes place following transcription. But the RNAi pathway is found to be absent in some eukaryotic protozoa. In case of *Leishmania major* and *Trypanosoma cruzi*, the components of RNAi are found to be missing (Robinson et al., 2003, DaRocha et al., 2004). The presence of RNA-mediated interference is depicted in *Saccharomyces cerevisiae* and some fungi (Aravind et al., 2000). The introduction of two RNAi-related proteins from *S. castellii* facilitated RNA interference in *S. cerevisiae* successfully (Drinnenberg et al., 2009). In certain ascomycetes and basidiomycetes the pathway is found to be missing, indicating that proteins required for RNA silencing have been lost from many fungal lineages in the course of evolution.

2.2. Mechanism of RNAi

Two kinds of RNA molecules, miRNA (micro RNA) and siRNA (small interfering RNA) play the central role in RNA interference. These small RNAs can bind to other specific messenger RNA (mRNA) molecules and increase or decrease their activity. MiRNA segments, each of approximately 21-27 nucleotides in length, encoding eukaryotic genomes are produced naturally. Each miRNA molecule is produced from a precursor transcript known as pre-miRNA, which is translocated from the nucleus into the cytoplasm and then cleaved into a mature miRNA (double stranded RNA molecules of 21-27 nts) by an enzyme known as Dicer. These miRNAs are unwound into two single-stranded RNAs (ssRNAs), popularly known as the passenger strand and the guide strand. The passenger strand is degraded while the guide strand gets incorporated into the RNA-induced silencing complex (RISC) which contains Argonaute protein having RNA cleavage activity. Guide strand present in the RISC complex directs it toward the complementary sequence in the target mRNA molecule for cleavage or translational inhibition, thus ultimately inhibiting protein synthesis (Figure 1).

Synthetic double stranded RNA (dsRNA) introduced into cells in culture and also into living organisms can selectively induce the suppression of specific genes of interest in the organism. This can be used to systematically knock down each gene in a cell and can thus help in the identification of components necessary for a particular cellular process or an event in a cell. The ability of a cell to uptake and incorporate foreign dsRNA varies from organism to organism. The effects of RNA interference may be both systemic and heritable in plants and animals. In plants, RNAi propagates by the transfer of siRNAs between cells through plasmodesmata. Heritability is achieved by the methylation of promoters targeted by RNAi. Specificity of the RNAi response lies in the process of targeting the endogenously produced miRNAs. In case of plants, miRNAs are usually known to bind complementarily to their target genes to induce cleavage of mRNA by the RISC complex, but in case of animals, miRNAs tend to become more divergent in sequence recognition and induce translational repression.

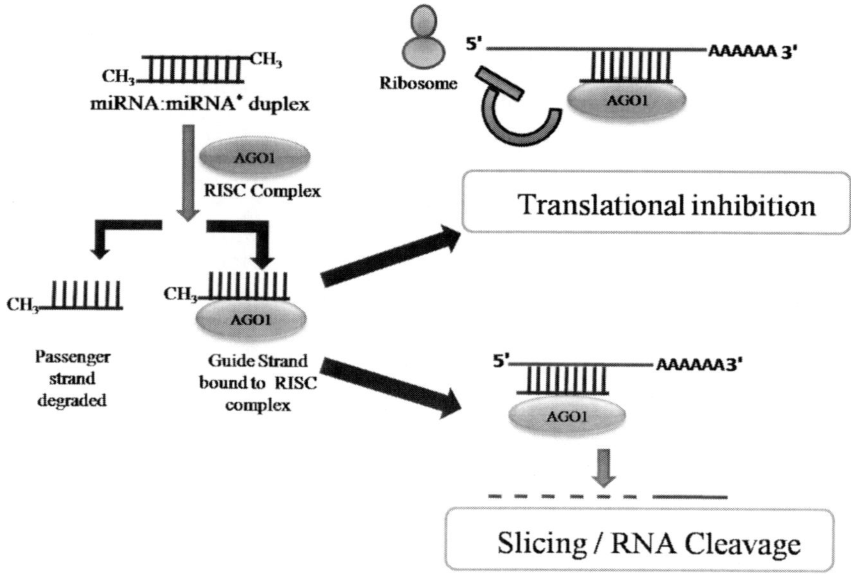

Figure 1. Mechanism of RNAi in plants.

The RNAi pathway is initiated when the dsRNA enters into the cytoplasm. The source of dsRNA may be: synthetic (siRNA) or single-stranded RNA (ssRNA) containing the two complementary sequences separated by a non-complementary sequence, that folds back on itself, forming a synthetic short hairpin RNA known as shRNA, expressed from a DNA construct which encodes a shRNA molecule. This is known as the dsRNAi approach. Synthetic approach involves the introduction of a molecule into the cytoplasm of the cell. In the dsRNAi approach, the cell itself produces the shRNA as a result of the DNA construct that is delivered to the nucleus and becomes part of the cells own DNA. The resultant gene silencing effect can last for months or years as the cell continues to produce its own DNA-directed shRNA to participate in the RNA interference pathway.

2.3. RNAi Latest Perspectives

RNAi is revolutionizing the way researchers study mammalian gene expression. RNAi is precise, efficient, stable and better than antisense technology for gene suppression. RNAi has had a significant impact on the ease, speed, and specificity with which the analysis of loss of gene function can be performed in mammalian cells and animal models. It has become the method of choice for researchers performing experiments on the loss of function studies in different organisms. The implications of this pathway are also being used as tools in biotechnology, medicine and insecticides. RNA interference plays a key role in helping the cells defend themselves against parasitic nucleotide sequences, viruses, and transposons and thus can influence development.

The human genome contains several genes that are candidates for drug targets. RNAi technology has been effectively used in target validation to identify and functionally assess relevant disease genes and can also be used as a tool to evaluate gene expression signatures in response to compounds or signalling pathways. RNAi is emerging as a powerful tool in animal models with applications being developed to use stable modified molecules such as RNAi or lentiviral delivery for effective in vivo studies.

Today we have a greater understanding of the components that are part of the RNAi pathway, the efficiency with which these components function in respect of sequence recognition and cellular mRNA cleavage and many of the key requirements for designing and generating extremely effective RNAi reagents. Our analysis of novel chemistries will further improve this approach and lead the way forward in in vivo analysis.

3. RNAi for Controlling Insects and Pests

Since its discovery, RNA interference has been used in different fields of molecular biology to determine gene functions, their role in the life cycle of an organism and in response to various biotic and abiotic stresses. RNAi has played an important role in decoding the vast amount of genetic information present in the genome sequence of an organism as it is practically possible to silence any gene of an organism by introducing or directing the synthesis of silencing molecules (dsRNA) against it. In insects, RNAi has gained a lot of success in both basic as well as applied research studies. In basic research, it is mainly used as a tool of reverse genetics to decipher the various components of the RNAi pathway and to understand the functions of genes in various physiological processes and developmental pathways. Silencing of some of the genes led to lethal phenotypes in insects, like delayed growth and development, developmental deformities such as inability of pupal formation or formation of defective pupa, and mortality. In the year 2007, two independent groups of researchers demonstrated the successful use of RNAi to control insect pests in plants. The first group, Baum and his coworkers developed transgenic maize plants resistant to western corn rootworm *Diabrotica virgifera virgifera* (order; Coleoptera) by using host-delivered RNAi technology (Baum et al., 2007). They silenced a mid-gut expressing gene coding for vacuolar ATPase enzyme in the rootworm. The second group led by Mao along with his co-workers developed transgenic *Arabidopsis* and tobacco plants against *Helicoverpa armigera* (order; Lepidoptera), by silencing its CYP6AE14 gene which codes for cytochrome P450 monoxygenase (Mao et al., 2007). Down regulation of CYP6AE14 of *H. armigera* increased its susceptibility towards gossypol, an insecticidal alkaloid present in cotton.

In recent years, several reports have been published elaborating the role of RNAi technology in achieving crop protection against the target insect-pests.

3.1. RNAi Delivery Methods in Insects

Different types of delivery methods are available for introducing dsRNA into the insect body.

3.1.1. Microinjection

Microinjection of dsRNA into the desired insect tissue is one of the most successful and widely used methods for gene functional studies. Among the various insect and pest species, dsRNA microinjection has been successfully used in the cut worm, *Spodoptera litura* (Rajagopla et al., 2002), grasshopper, *Schistocerca americana* (Dong and Friedrich, 2005), pea aphid, *Acyrthosiphon pisum* (Mutti et al., 2006), and plant hopper, *Nilaparvata leugens* (Liu et al., 2010). However, application of microinjection to control insect pests at field level is not a feasible approach because of its intrinsic technical limitations. It is a highly laborious and skillful process, and moreover a practically impossible task to pick individual insects from the field and perform microinjection. The process of microinjection is invasive in nature and causes non specific mechanical damage to the insect species, and therefore is not preferred when targeting embryo, soft and small bodied insects. Several factors decide the efficiency of microinjection for causing RNAi like injection volume, site of injection (accessible or not), and the location of target tissue.

3.1.2. Artificial Diet Feeding

In the case of artificial diet feeding, bacterial cells expressing dsRNA molecules or chemically synthesised or *in vitro* transcribed dsRNA molecules are mixed in the artificial diet. DsRNA present in the diet reaches inside insect body upon feeding. Although artificial diet feeding is a non invasive method and technically less demanding as compared to microinjection, it is cumbersome to perform as fresh diet has to be prepared regularly for the purpose of feeding. Moreover, *in vitro* transcription and bacterial cloning requires extra efforts and resources

(Kumar et al., 2009; Shakesby et al., 2009; Mao et al., 2015; Zhang et al., 2015).

3.1.3. Soaking

Soaking as a method of dsRNA delivery was first used in the case of nematodes followed by insects. In this method, target insects are soaked in a solution containing dsRNA molecules of the target gene. To execute successful RNAi through soaking, these double stranded RNA molecules must be able to pass through the insect body wall, stoma and inter-segmental membrane. Like above methods, soaking method also suffers from technical limitations, and cannot be applied at the field level (Aronstein et al., 2006).

3.1.4. Spraying or Topical Application

Spraying of dsRNA at the field level is similar to the application of chemical insecticides. Like soaking, spraying also requires absorption of dsRNA through the body wall to manifest RNAi. Spraying may become a viable approach for insects and pests control if large amount of dsRNA can be produced cheaply as it is able to reduce pest population faster than the conventional insecticides (Gan et al., 2010; Wang et al., 2011).

In microinjection method, stability of dsRNA is mainly governed by the internal environment of the tissue or the site where dsRNA molecules are injected whereas in artificial diet feeding, spraying and soaking methods, stability of dsRNA molecules is governed by both external (outside the insect body) as well as internal (inside the insect body) environments, as in these methods dsRNA remains in the external environment for a sufficient period of time. Some of the basic research studies, involving artificial diet feeding showed that for efficient RNAi in insects, it is mandatory to have a continuous supply of dsRNA as few doses of dsRNA can lead to transient silencing of the specific gene that would lead to the recovery of the insect from the silencing effect (Asokan et al., 2014; Zhang et al., 2015). This transient silencing is due to the absence of RNAi amplification machinery in insect species which is otherwise present in the case of nematodes and responsible for increasing the amount of initial silencing signal in a substantial amount (Sijen et al., 2001; Tomoyasu et al., 2008). Thus, for controlling of insect and pest population at field level, a suitable delivery method is needed which is autonomous or self sustaining and acts as a continuous supplier of dsRNA to the concerned insect pest. This requirement led to the development of host delivered RNAi technology (HD-RNAi) or RNAi through transgenic plants.

Table 1. Development of insects-pest resistant transgenic plants through host-delivered RNAi technology

Target insect species	Target gene or Protein silenced through RNAi	Crop	Bioassay observations	References
Helicoverpa armigera	Cytochrome P450 gene (*CYP6AE14*)	*Arabidopsis* &Tobacco	Larval growth retardation	Mao et al., 2007
Diabrotica virgifera	*vATPase A*	Maize	Larval stunting and mortality	Baum et al., 2007
Nilaparvata leugens	Mid-gut genes	Tobacco	Reduction in transcripts of target genes, no lethal phenotypic effect	Zha et al., 2011
Myzus persicae	Serine Protease	*Arabidopsis*	Reduced fecundity	Bhatia et al., 2012
Helicoverpa armigera	Ecdysone receptor gene (*EcR*)	Tobacco	Moulting defects & larval mortality	Zhu et al., 2012
Helicoverpa armigera	*HR3*	Tobacco	Larval mortality and developmental deformities and	Xiong et al., 2013
Myzus persicae	*(V-ATPaseE), tubulin folding cofactor D (TBCD), AChE2*	Tobacco	Transgenic plants showed resistance towards aphid	Guo et al., 2014
Helicoverpa armigera	Arginine Kinase (*AK*)	*Arabidopsis*	Larval mortality	Liu et al., 2015
Helicoverpa armigera	HMGCoA reductase; *HMGR*	Cotton	Larval growth retardation	Tian et al., 2015
Helicoverpa armigera	*Chitinase*	Tomato and Tobacco	Larval mortality, growth retardation, and developmental deformities	Mamta et al., 2016

3.1.5. Host-Delivered RNAi Technology (HD-RNAi)

In host-delivered RNAi technology, host plants are genetically engineered to produce dsRNA molecules specific to the essential gene of the target pests. Whenever insects and pests feed on such host plants, dsRNA molecules expressing in them are transferred into the insect's body to initiate RNA silencing. Thus a continuous supply of dsRNA takes place from the host plant to the insect upon feeding. Moreover, as the genetically engineered host plant is a natural food for the target pest, feeding is autonomous and does not require any delivery method. In recent years, several successful examples of HD-RNAi technology have been reported (Table 1).

3.2. Genetic Engineering Strategies for the Production of dsRNA Molecules in the Host Plant

Genetic engineering of host plants for the production of dsRNA precursor molecules is mainly carried out by two different approaches:

3.2.1. Hair-Pin RNAi Strategy

In the case of hair-pin RNAi strategy, the vector construct contains both the sense and antisense strands of the target gene together, separated by a spacer sequence (an intronic sequence). Upon transcription, precursor RNA molecule forms a perfect hair-pin stem loop structure. Stem region is formed by the Watson and Crick base pairing between the transcribed sense and antisense strands whereas loop region is formed by the intronic sequence. This precursor molecule is recognised by the plant RNAi machinery. Upon recognition, the precursor RNA molecule is processed in a manner similar to the natural precursors of siRNA molecules. After processing single precursor molecule gives rise to a pool of duplex siRNA molecules (21-24nts) from its stem region, each having a different sequence. Due to the synthesis of a pool of siRNA molecules, chances of off-target and non-target effects are higher in this case; off-target effects dilute the silencing of the targeted gene as generated siRNA molecules also cause silencing of unintended genes in the organism whereas non-target effects lead to the silencing of genes in other non-target organisms. Thus hairpin RNAi based strategy raises ecological concerns and biosafety issues prior to their release into nature (Tiwari et al., 2014). Different cloning strategies for the synthesis of hair-pin RNAi construct include: restriction and ligation based vector method, direct amplification of intron containing hair-pin (Xiao et al., 2006), overlap-extension PCR and golden gate cloning based restriction –ligation method (Yan et al., 2012).

3.2.2. Artificial microRNA Based Strategy

In the case of artificial microRNA based strategy, the vector construct is prepared by replacing the original microRNA/microRNA* duplex of natural plant microRNA with the artificial microRNA/microRNA* sequence without altering its backbone (Schwab et al., 2006). Upon transcription, precursor RNA molecule forms an imperfect hair-pin stem loop structure, a typical structure of a natural miRNA precursor molecule. This precursor molecule is recognised by a plant microRNA generating machinery and processed in a manner similar to natural miRNA precursor molecule. After processing single miRNA precursor molecule gives rise to single pre-designed amiRNA molecule of 21nts. In this strategy, single artificial microRNA is generated from the precursor molecule thus is more specific in its silencing and has less off-target and non target effects (Tiwari et al., 2014). Designing of artificial microRNA requires use of different bioinformatics tools like WMD (http://wmd3.weigelworld.org/cgi-bin/webapp.cgi), Ps RNA target (http://plantgrn.noble.org/psRNATarget/), Viena RNA package (http://www.tbi.univie.ac.at/RNA/) etc. After obtaining potential amiRNAs against the target gene, these amiRNAs are first screened on the basis of the most common determinants of plant microRNAs (U at 5′ end, A at position 10, and preferentially C at position 19 and UU at 3′ end) and the amiRNAs obtained are further screened for any off-targets and non-target effects against human and beneficial insects by searching for their complementary sequence in the available sequence databases. The amiRNA sequences which do not show any complementarity in the searched database are finally selected.

Different cloning strategies to prepare amiRNA based constructs are uracil excision based cloning strategy (Ossowaki et al., 2008), simple amiRNA vector designing method (Wang et al., 2012), two step designing method (Zhou et al., 2013), universal amiRNA based vectors (Zhou et al., 2013), seamless enzyme–free cloning (SEFC) and mating assisted genetically integrated cloning (MAGIC) method (Yan et al., 2011).

3.3. RNAi Response in Insects

RNAi response in insects is broadly classified into two classes, namely, cell autonomous RNAi and non cell autonomous RNAi. Non cell autonomous RNAi is further classified into environmental RNAi and systemic RNAi (Whangbo and Hunter, 2008). In cell autonomous RNAi, silencing effect is restricted to the particular cell where dsRNA is either endogenously synthesised or introduced exogenously whereas in environmental RNAi, RNA silencing signal is taken up by the cells from the immediate extracellular environment such as gut or haemocoel or from surroundings through skin. In systemic RNAi, silencing signal moves from its site of uptake to the neighbouring or distantly placed cells or tissues.

Systemic RNAi becomes important when the target gene expression takes place in a tissue which lies beyond the mid-gut tissue. For systemic RNAi, environmental RNAi is the pre-requisite (Whangbo and Hunter, 2008). For the successful application of Host-delivered RNAi technology in crop protection against insects and pests, non-cell-autonomous RNAi is must. In the case of *C. elegans* systemic RNAi is associated with amplification of RNAi through RNA dependent RNA polymerase (RdRP) but in the insects till now no RdRP homolog of *C. elegans* has been discovered, suggesting either the presence of an alternate mechanism of RNAi amplification or its complete absence in insects (Tomoyasu et al., 2008). Due to this limitation, continuous delivery of dsRNA molecules through the host plant is a must to bring desirable silencing effects in insects.

3.4. Different Factors Responsible for Efficient HD-RNAi

RNAi application and efficacy varies with the target genes, tissues, insect species and their life stages (Terenius et al., 2011). Terenius et al, 2011 have discussed several successful and unsuccessful experiments of RNAi with respect to lepidopteran insect species. Out of 130 selected target genes only 38% were silenced at satisfactory levels, while 14% were showed insufficient silencing and remaining 48% did not show any silencing. Some of the critical factors which need to be considered while employing RNAi for insects and pests control are discussed below.

3.4.1. Choice of the Target Gene

Selection of the target gene should be carefully done especially in cases where the ultimate objective is pest control. Important factors while the selection of target gene include:

3.4.1.1. Function of the Target Gene

A target gene must code for an essential function in the organism; silencing of which leads to lethal phenotypic effects, developmental deformities or reduced fecundity, which ultimately results in the reduction of insects and pests population in the coming generations (Zhu et al., 2012; Xiong et al., 2013; Asokan et al., 2014; Liu et al., 2015; Tian et al., 2015; Jin et al., 2015; Lim et al., 2016).

3.4.1.2. Site of the Target Gene Expression

Efficiency of RNAi has been found to vary among different tissues in an insect. Especially in the case of oral delivery methods, genes expressing in the mid-gut cells are the most preferred targets for RNAi because silencing signals have to travel less distance before reaching their target tissue which is mid-gut in this case (Parrish et al., 2000; Mao et al., 2011; Jayachandran et al., 2013; Mao et al., 2015). Less distance results in less physical and chemical barriers to cross thus, results in better RNAi response. However for successful RNAi in target genes expressing in the tissues beyond mid-gut region, higher optimum concentrations of dsRNA are needed to compensate for the losses due to above mentioned barriers.

3.4.1.3. Half Life of the Protein

Final result of post transcriptional gene silencing is the inhibition of protein synthesis, which is either mediated by the cleavage of concerned mRNA molecule or by repressing its translation into a protein molecule. Half life of a protein is an important indicator of its stability. Silencing of genes which code for more stable protein molecules in the cell is difficult and they exhibit very poor silencing for example, nicotinic acetylcholine receptors (nAChRs) can remain stable for more than two weeks (Lomazzo et al., 2011), and this was responsible for its weak RNAi response in both *T. castaneum* and *D. melanogaster* (Rinkevic and Scott, 2013). Thus before selecting a particular gene as a target for RNAi, preliminary information about the half life of its encoded protein must be investigated or known.

3.4.1.4. Transcript Abundance

Ideally chances of silencing a gene whose transcript levels are low and the encoded protein has a short half life, are more. Genes having lower amount of transcripts require lesser amount of dsRNA for their silencing (Asokan et al., 2014).

3.4.1.5. Target Gene Region for the Synthesis of dsRNA Molecule or amiRNA Molecule

Transcription unit of a gene comprises of coding and non coding regions. Exonic sequences form the coding region whereas 5' and 3' untranslated regions and intronic sequences form non coding regions of the gene. For the synthesis of dsRNA molecules, selection of UTR and exonic sequences is more common. In case of genes which are highly conserved among the various insects and pests or which belong to a gene family within a particular insect-pest, designing of dsRNA from the conserved region will be able to target more than one pest or genes simultaneously (Zhu et al., 2012).

3.4.2. dsRNA Concentrations in the Host Plants

For effective RNAi, an optimal concentration of dsRNA is required in the entire host plant or in the desired plant tissue which is preferred by the insect for feeding, for example to control the insect-pest which feeds on the phloem sap, expression of dsRNA is desirable in the phloem tissue under phloem specific promoter (Upadhyay et al., 2011).

3.4.3. Length of the dsRNA Fragment

Length of the dsRNA fragment is an important indicator of uptake efficiency in insects and therefore plays a key role in silencing of a target gene. In case of transgenic plant mediated RNAi, long unprocessed hair-pin dsRNAs are found to have more pronounced silencing effect as compared to host plant processed dsRNA (Mao et al., 2007; Kumar et al., 2012). In few recent studies, dsRNA production in the chloroplast resulted in stronger silencing effects in the target insect as compared to its expression in the nucleus. Since chloroplast lacks RNAi machinery, dsRNA formed after transcription of the hair-pin RNAi construct remains unprocessed and leads to better silencing over nuclear processed dsRNA precursors (Zhang et al., 2015; Bally et al., 2016). In contrast, other studies showed efficient silencing of target gene in insects and pests by 21 nucleotide amiRNA through both host delivered RNAi and artificial diet feeding (Agrawal et al., 2015; Yogindran et al., 2016).

3.4.4. Alkaline Nature of the Gut

Most of the insect species have alkaline pH in their gut lumen. Presence of alkaline conditions favours chemical hydrolysis of dsRNA molecules into individual ribonucleotides, thus destroying their silencing behaviour (Dow, 1992; Terra et al, 1996).

3.4.5. RNase Present in the Digestive Juice and Haemolymph

Presence of non-specific RNA degrading enzymes in the gastric juice and haemolymph across several insects order pose another barrier against effective RNAi (Arimatsu et al., 2007; Liu et al., 2012; Garbutt et al.,

2013;Yang and Han, 2014; Christiaens et al., 2014; Wang et al., 2016; Shukla et al., 2016).

3.4.6. RNA Uptake Mechanism in Insects, Still a Mystery to Be Resolved

Molecular mechanisms by which mid-gut epithelial cells take up dsRNA from the gut lumen are not very well known in most of the insect species. Preliminary information about the uptake mechanisms is available from *Drosophila* S2 cell lines only (Saleh et al., 2006; Ulvila et al., 2006; Jose and Hunter, 2007). DsRNA uptake mechanisms are still mysterious even for a highly sensitive species such as western corn rootworm and *Tribolium* (Tomoyasu et al., 2008; Xu and Han, 2008; Zha et al., 2011; Shukla et al., 2016).

3.5. Prospects of RNAi-Based Insect-Pest Control

RNAi technology has gained a lot of success in the development of insect and pest tolerant crops. Specificity of RNAi against the target insect-pests while leaving other species unaffected is the most important ecological factor which needs to be considered prior to their release for commercial cultivation. RNAi is equally effective against various types of insect–pests, including piercing and sucking type of mouth-parts unlike Bt-technology which is mainly restricted to insect-pests having cutting and chewing type of mouth parts. Several bioinformatics softwares are available which help in the designing of more accurate silencing signals, having minimum off-target effects. Other potential applications of RNAi in pest control include disruption of communication system of insects by silencing the receptors which are involved in communication relay such as silencing of pheromone receptors causing difficulty in searching the appropriate mating partners (Turner et al., 2006; Trivedi, 2010; Zhao et al., 2011; Zhang et al., 2012). By decreasing the expression of insecticide resistant genes in the concerned insect species through RNAi, susceptibility of insects and pests can be improved for the same. It can also decrease the usage of particular insecticide. Thus RNAi can serve as a more eco-friendly approach for pest control (Bautista et al., 2009; Zhang et al., 2012).

4. Application of RNA Interference in Crop Improvement

Besides having a role in crop protection, RNAi is used in several other areas of crop improvement. It is successfully applied in various animal model systems including human cell lines to study intricacies of disease development and can also be used as a therapeutic tool for curing several diseases.

4.1. In Controlling Developmental Timings

Several miRNA families have been revealed to play central roles in a number of the pathways controlling flowering, serving either to hold back or to promote the reproductive phase transition. The main players are the *miR156* and *miR172* families, the activities of which control both the juvenile to adult vegetative phase transition and reproductive phase transition. In *Arabidopsis.* FLC gene has been reported to delay flowering time. For early flowering, FLC can be silenced through RNAi.

4.2. Abiotic Stress Tolerance

Most of the environmental stresses such as drought, flooding, salinity and temperature, reduce the productivity of plants by a great extent. Some of the recent evidence indicates involvement of RNAi in stress responses in plants. Up-regulation of microRNA393 is reported in *Arabdopsis* during cold, dehydration, high salinity and abscisic acid treatment. It is also seen that there is an up regulation of miR395 and miR399 gene during sulphur and phosphate starvation (Rhoades and Bartel, 2004). Thus, modifying expression of such microRNAs through transgenic approach can help in the generation of climate resilient crops.

4.3. Biotic Stress Tolerance

Biotic stress factors like parasitic weeds, nematodes, insects and fungal diseases cause severe damage to the crop yield, affecting total agricultural production. In this chapter the role of RNAi in crop protection against insect-pests has been already described in detail. RNAi has been successfully used to generate virus resistant plants, because viruses are very difficult to control through available strategies.

4.4. Virus Resistance

It has been reported that plant activates RNAi machinery against RNA plant viruses. RNA genome of such plant viruses replicates via double stranded RNA intermediates and acts as a trigger for RNAi in plants. It is confirmed that plants defective in RNAi show hypersensitivity to virus infection. Transgenic blackgram (*Vigna mungo*) plants, expressing dsRNA against the promoter sequence of geminivirus (VMYMV) under the control of the CaMV35S promoter showed resistance against *Vigna mungo* yellow mosaic virus (VMYMV) (Duan et al., 2012). Potato *(Solanum tuberosum)* plants are very susceptible to potato virus X family *(Alphaflexi viridae)*. RNAi technology has been successfully used to impart resistance in potato against potato virus X. A commercial variety of potato (cv. Spunta) transformed with RNAi construct expressing dsRNA against the coat protein region of PVY showed resistance against three different strains of PVY. Banana is the major crop of South East Asia and India is very much prone to the banana bract mosaic virus (BBrMV), mainly transferred by aphids. Resistance was achieved in banana plants by expressing double stranded RNA against coat protein of the virus through RNAi.

4.5. Fungal Resistance

Aspergillus, Fusarium, and Penicillium fungal species are the major source of mycotoxins such as aflatoxins (cancer causing agent), 4-deoxynivalenol (inhibitor of protein synthesis), and patulin (immunotoxic) in food and feed crops. Besides conventional approaches to control fungal attack on food crops, biotechnologists have adopted HIGS (Host Induced Gene Silencing), as a major gene silencing method for controlling mycotoxic attack in crop plants (Majumdar et al., 2017).

RNAi mediated resistance against *Phytophthora parastica* by targeting glutathione s-transferase gene in *Nicotiana tabaccum* has resulted in significant increase in resistance to infection.

4.6. Nematode Resistance

RNAi tools can be used to control the damage caused by the root knot nematodes (RKN) (*Meloidogyne* spp.) which cause large economic losses worldwide. Root knot nematodes reside inside the soil and drain plant nutrients. Use of nematicides to control RKNs has been banned due to their high toxicity and ozone-depleting properties, thus there is high demand for an alternative control to combat nematodes. Transgenic tobacco (*Nicotiana tabacum*) plants producing dsRNA targeting parasitic genes and transcription factors of RKNs have shown higher level of resistance against control plants (Tamilarasan et al., 2013).

4.7. Resistance to Bacterial Diseases

RNAi is also used as a prominent tool to protect plants against pathogenic bacteria, mainly *Agrobacterium tumefaciens* which causes crown gall disease. RNAi technology can be used to silence iaaM and ipt gene to induce resistance against crown gall (Jagtap et al., 2011).

4.8. Development of Male Sterile Plants: Hybrid Seed Production

RNAi is used to develop male sterile plants which can be exploited in hybrid seed industry. Engineering male sterility is especially useful in crop species where natural male sterile lines are not available. Genes which are specific to pollen production can be silenced through RNAi. For instance, scientists have developed male sterile tobacco by silencing the expression of TA29, a gene involved in pollen development.

4.9. Improvement of Nutritional Value

Malnutrition has affected more than 40% of world with respect to micronutrients and vitamins. RNAi can be used for the improvement of the nutritional value in potato by silencing beta–carotene hydroxylase gene which enhances accumulation of beta-carotene. Similarly in the seeds of T3 generation transgenic *Brassica napus* plants, sinapate esters levels were reduced by 76% through RNAi by silencing UDP- Glc: sinapateglucosyl transferase gene activity (Jagtap et al., 2007).

4.10. Removal of Toxic Compounds

Most plants contain toxins of various types, which create problems in extraction of desirable products. RNAi can be used to minimise the content of toxic compounds.

Cotton: Cotton seeds are rich in protein, and can be used as protein source in diet, but presence of gossypol, a toxic compound makes it unfit for human consumption. Enzyme delta-cadinene synthase (DCS) is involved in the synthesis of gossypol. With the help of RNAi, enzymatic gene DCS can be silenced in the seed tissues without affecting its production in the other parts of the plants as it has defensive role against plant pathogens (Rathore et al., 2011).

Coffee: Most of the world prefers to drink decaffeinated coffee which is low in caffeine. Caffeine is very energising for the central nervous system, respiratory and cardiac system. Its adverse side effects include insomnia, restlessness and palpitations. RNAi can be used to produce natural coffee which contains a low amount of caffeine (Pathak et al., 2015).

CONCLUSION

RNAi technology holds a great potential for the management of pests attacking agricultural crops. Its specific mode of action and wide applicability against a variety of insects and pests makes it an ideal choice for achieving crop protection. Among the various delivery methods discussed here, host–delivered RNAi technology seems to be more promising and has a tremendous potential for the management of insects at the field level because of its self sustaining and ecofriendly nature. Although several examples of HD-RNAi technology have been reported for the development of insect tolerant crops, there are still many limitations associated with it which need to be discussed and resolved for realising its full potential. Apart from the selection of target gene and delivery methods, expression of double stranded RNA in the host plant is an important factor to be considered. Stability of dsRNA or dsRNA derived small RNA depends on several factors within the insect, like gut pH, amount of RNA degrading enzymes in the gastric juice and haemolymph, and uptake efficiency. An optimal concentration of dsRNA needs to be expressed in the host plant to have effective RNAi. Questions about the optimal concentration of dsRNA required for a particular pest, expression levels, specificity of the tissues need to be answered. Along with the expression strategies, risks associated with the development of resistance in insects and pests against RNAi causing agent also needs an evaluation. Targeting different regions of the same gene or different genes in the same organism as a combinatorial approach may delay development of resistance against RNAi and offer better pest control. Thus, RNAi technology can be used as an alternative tool along with the existing *Bt*-technology for controlling agricultural pests, provided its molecular mechanisms along with its limitations in the target pest are well known.

REFERENCES

Agrawal, A., Rajamani, V., Reddy, V. S., Mukherjee, S. K. and Bhatnagar, R. K. (2015). Transgenic plants over-expressing insect-specific microRNA acquire insecticidal activity against *Helicoverpa armigera*: an alternative to Bt-toxin technology. *Transgenic Research*, 24:791-801.

Allison H. S. Hall and Kenneth A. Alexander. (2003). RNA Interference of Human Papillomavirus Type 18 E6 and E7 induces senescence in HeLa Cells. *Journal of Virology*, 6066–6069.

Arimatsu, Y., Kotani, E., Sugimura, Y. and Furusawa, T. (2007). Molecular characterization of a cDNA encoding extracellular dsRNase and its expression in the silkworm, *Bombyx mori*. *Insect Biochemistry and Molecular Biology*, 37:176–183.

Aronstein, K., Pankiw, T. and Saldivar, E. (2006). SID-1 is implicated in systemic gene silencing in the honey bee. *Journal of Apicultural Research*, 45:20–24.

Asokan, R., Chandra, G. S., Manamohan, M., Kumar, N. K. K. and Sita, T. (2014). Response of various target genes to diet-delivered dsRNA mediated RNA interference in the cotton bollworm, *Helicoverpa armigera*. *Journal of Pest Science*, 87: 163–172.

Bally, J., McIntyre, G. J., Doran, R. L., Lee, K., Perez, A., Jung, H., Naim, F., Larrinua, I. M., Narva, K. E. and Waterhouse, P. M. (2016). In-Plant Protection against Helicoverpa armigera by Production of Long hpRNA in Chloroplasts. *Frontiers in Plant Science*, 7:1453.

Baum, J. A., Bogaert, T., Clinton, W., Heck, G. R., Feldmann, P., Ilagan, O., Johnson, S., Plaetinck, G., Munyikwa, T., Pleau, M., Vaughn, T. and Roberts, J. (2007). Control of coleopteran insect pests through RNA interference. *Nature Biotechnology*, 25:1322–1326.

Bautista, Ma. A. M., Miyata, T., Miura, K. and Tanaka, T. (2009). RNA interference-mediated knockdown of a Cytochrome P450, CYP6BG1, from the diamondback moth, *Plutella xylostella*, reduces larval resistance to permethrin. *Insect Biochemistry and Molecular Biology*, 39:38–46.

Bhatia, V., Bhattacharya, R., Uniyal, P. L., Singh, R. and Niranjan R. S. (2012) Host generated siRNAs attenuate expression of serine protease gene in *Myzus persicae*. *Plos One*, 7: e46343.

Christiaens, O., Sweveres, L. and Smagghe, G. (2014). DsRNA degradation in the pea aphid (Acyrthosiphon pisum) associated with lack of response in RNAi feeding and injection assay. *Peptides*, 53:307–314.

Delmas, O., Durand-Schneider, A. M., Cohen, J., Colard, O and Trugnan, G. (2004). Spike Protein VP4 Assembly with maturing rotavirus requires a postendoplasmic reticulum event in polarized caco-2 Cells. *Journal of Virology*, 10987–10994.

Dong, Y. and Friedrich, M. (2005). Nymphal RNAi: systemic RNAi mediated gene knockdown in juvenile grasshopper. *BMC Biotechnology*, 5: 25.

Dow, J. (1992). pH gradients in lepidopteran midgut. *Journal of Experimental Biology*, 172(1): 355–375.

Gan, D., Zhang, J., Jiang, H., Jiang, T., Zhu, S. and Cheng, B. (2010). Bacterially expressed dsRNA protects maize against SCMV infection. *Plant Cell Reports*, 29:1261–1268.

Garbutt, J. S., Bellés, X., Richards, E. H. and Reynolds, S. E. (2013). Persistence of double-stranded RNA in insect hemolymph as a potential determiner of RNA interference success: evidence from *Manduca sexta* and *Blattella germanica. Journal of Insect Physiology*, 59:171–178.

Guo, H., Song, X., Wang, G., Yang, K., Wang, Y., Niu, L., Chen, X. and Fang, R. (2014) Plant-generated artificial small RNAs mediated aphid resistance. *PLoS One*, 9(5): e97410.

Jayachandran, B., Hussain, M. and Asgari, S. (2013). An insect trypsin-like serine protease as a target of microRNA: utilization of microRNA mimics and inhibitors by oral feeding. *Insect Biochemistry and Molecular Biology*, 43:398-406.

Jin, S., Singh, N. D., Li, L., Zhang, X. and Daniell, H. (2015). Engineered chloroplast dsRNA silences cytochrome p450 monooxygenase, V-ATPase and chitin synthase genes in the insect gut and disrupts *Helicoverpa armigera* larval development and pupation. *Plant Biotechnology Journal*, 13:435–446.

Jones-Rhoades, MW, Bartel, DP (2004). Computational identification of plant microRNAs and their targets, including a stress-induced miRNA. *Mol Cell,* 18; 14(6):787-99.

Jose, A. M. and Hunter, C. P. (2007). Transport of sequence-specific RNA interference information between cells. *Annual Review of Genetics*, 41:305–330.

Kumar, M., Gupta, G. P. and Rajam, M. V. (2009). Silencing of acetylcholinesterase gene of *Helicoverpa armigera* by siRNA affects larval growth and its life cycle. *Journal of Insect Physiology*, 55:273–278.

Kumar, P., Pandit, S. S. and Baldwin, I. T. (2012). Tobacco Rattle Virus Vector: A rapid and transient means of silencing *Manduca sexta* genes by plant mediated RNA interference. *PLoS One.*

Lim, Z. X., Robinson, K. E., Jain, R. G., Sharath Chandra, G., Asokan, R., Asgari, S. and Mitter, N. (2016). Diet-delivered RNAi in *Helicoverpa armigera* - progresses and challenges. *Journal of Insect Physiology*, 85:86-93.

Liu, F., Wang, X., Zhao, Y., Li, Y., Liu, Y. and Sun, J. (2015). Silencing the HaAK gene by transgenic plant-mediated RNAi impairs larval growth of *Helicoverpa armigera. International Journal of Biological Sciences*, 11:67–74.

Liu, J., Swevers, L., Iatrou, K., Huvenne, H. and Smagghe, G. (2012). *Bombyx mori* DNA/RNA non-specific nuclease: expression of isoforms in insect culture cells, subcellular localization and functional assays. *Journal of Insect Physiology*, 58:1166–1176.

Liu, S., Ding, Z., Zhang, C., Yang, B. and Liu, Z. (2010). Gene knockdown by intro-thoracic injection of double-stranded RNA in the brown planthopper, *Nilaparvata lugens. Insect Biochemistry and Molecular Biology*, 40:666–671.

Lomazzo, E., Hussmann, G. P., Wolfe, B. B., Yasuda, R. P., Perry, D. C. and Kellar, K. J. (2011). Effects of chronic nicotine on heteromeric neuronal nicotinic receptors in rat primary cultured neurons. *Journal of Neurochemistry,* 119: 153–164.

Majumdar, R., Rajasekaran, K., and Cary, J. W. (2017). RNA Interference (RNAi) as a potential tool for control of mycotoxin contamination in crop plants: concepts and considerations. *Frontiers in Plant Science*, 8: 200.

Mamta, Reddy, K. R. and Rajam, M. V. (2016). Targeting chitinase gene of *Helicoverpa armigera* by host-induced RNA interference confers insect resistance in tobacco and tomato. *Plant Molecular Biology*, 90:281–292.

Mao, J., Zhang, P., Liu, C. and Zeng, F. (2015). Co-silence of the coatomerβ and v-ATPase A genes by siRNA feeding reduces larval survival rate and weight gain of cotton bollworm, *Helicoverpa armigera. Pesticide Biochemistry and Physiology*, 118:71–76.

Mao, Y. B., Cai, W. J., Wang, J. W., Hong, G. J., Tao, X. Y., Wang, L. J., Huang, Y. P. and Chen, X. Y. (2007). Silencing a cotton bollworm

P450 monooxygenase gene by plant-mediated RNAi impairs larval tolerance of gossypol. *Nature Biotechnology*, 25:1307–1313.

Mao, Y. B., Tao, X. Y., Xue, X. Y., Wang, L. J. and Chen, X. Y. (2011) Cotton plants expressing *CYP6AE14* double-stranded RNA show enhanced resistance to bollworms. *Transgenic Research*, 20: 665–673.

Martineau, H. M and Pyrah, T. H. (2017). Review of the application of RNA interference technology in the pharmaceutical industry. *Toxicologic Pathology*, 35:327–336.

Mutti, N. S., Park, Y., Reese, J. C. and Reeck, G. (2006). RNAi knockdown of a salivary transcript leading to lethality in the pea aphid, *Acyrthosiphon pisum*. *Journal of Insect Science*, 6:1–7.

Ossowski, S., Schwab, R. and Weigel, D. (2008) Gene silencing in plants using artificial microRNAs and other small RNAs. *Plant Journal*, 53:674–690

Palle, R. S., Campbell, L. M., Pandeya, D., Puckhaber, L., Sabarinath Sundaram, S., Rathore, K. S. (2012). RNAi-mediated ultra-low gossypol cottonseed trait: performance of transgenic lines under field conditions. *Plant Biotechnology Journal*, 11: 296–304.

Parrish, S., Fleenor, J., Xu, S., Mello, C. and Fire, A. (2000). Functional anatomy of a dsRNA trigger: differential requirement for the two trigger strands in RNA interference. *Molecular Cell*, 6:1077–1087.

Pathak, K. and Gogoi, B. (2015). RNA interference (RNAi): application in crop improvement: a review. *Agricultural Reviews*, 37 (3): 245-249.

Rajagopal, R., Sivakumar, S., Agrawal, N., Malhotra, P. and Bhatnagar, R. K. (2002). Silencing of midgut aminopeptidase N of *Spodoptera litura* by double-stranded RNA establishes its role as *Bacillus thuringiensis* toxin receptor. *Journal of Biological Chemistry*, 277:46849–46851.

Rinkevich, F. D. and Scott, J. G. (2013). Limitations of RNAi of α6 nicotinic acetylcholine receptor subunits for assessing the *in vivo* sensitivity to spinosad. *Insect Science*, 20: 101–108.

Saleh, M. C., van Rij, R. P., Hekele, A., Gillis, A., Foley, E., O'Farrell, P. H. and Andino, R. (2006). The endocytic pathway mediates cell entry of dsRNA to induce RNAi silencing. *Nature Cell Biology*, 8:793–802.

Schwab, R., Ossowski, S., Riester, M., Warthmann, N. and Weigel, D. (2006). Highly specific gene silencing by artificial microRNAs in *Arabidopsis*. *Plant Cell*, 18:1121-1133.

Shakesby, A. J., Wallace, I. S., Isaacs, H. V., Pritchard, J., Roberts, D. M. and Douglas, A. E. (2009). A water-specific aquaporin involved in aphid osmoregulation. *Insect Biochemistry and Molecular Biology*, 39:1–10.

Shukla, J. N., Kalsi M., Sethi, A., Narva, K. E., Fishilevich, E., Singh, S., Mogilicherla, K and Palli, S. R. (2016). Reduced stability and intracellular transport of dsRNA contribute to poor RNAi response in lepidopteran insects. *RNA Biology*, 13(7): 656-669.

Sijen, T., Fleenor, J., Simmer, F., Thijssen, K. L., Parrish, S., Timmons, L. (2001). On the role of RNA amplification in dsRNA-triggered gene silencing. *Cell*, 107:465-76.

Tamilarasan, S., and Rajam, M. V. (2013). Engineering crop plants for nematode resistance through Host-Derived RNA Interference. *Cell and Developmental Biology*, 2:2.

Tan, F. L., James, Q. and Yin, J. Q. (2004). RNAi, a new therapeutic strategy against viral infection. *Cell Research*, 14(6):460-466.

Terenius, O., Papanicolaou, A., Garbutt, J. S., Eleftherianos, I., Huvenne, H., Kanginakudru, S., Albrechtsen, M., An, C., Aymeric, J.-L., Barthel, A., Bebas, P., Bitra, K., Bravo, A., Chevalier, F., Collinge, D. P., Crava, C. M., de Maagd, R. A., Duvic, B., Erlandson, M., Faye, I., Felföldi, G., Fujiwara, H., Futahashi, R., Gandhe, A. S., Gatehouse, H. S., Gatehouse, L. N., Giebultowicz, J. M., Gómez, I., Grimmelikhuijzen, C. J. P., Groot, A. T., Hauser, F., Heckel, D. G., Hegedus, D. D., Hrycaj, S., Huang, L., Hull, J. J., Iatrou, K., Iga, M., Kanost, M. R., Kotwica, J., Li, C., Li, J., Liu, J., Lundmark, M., Matsumoto, S., Meyering-Vos, M., Millichap, P. J., Monteiro, A., Mrinal, N., Niimi, T., Nowara, D., Ohnishi, A., Oostra, V., Ozaki, K., Papakonstantinou, M., Popadic, A., Rajam, M. V., Saenko, S., Simpson, R. M., Soberón, M., Strand, M. R., Tomita, S., Toprak, U., Wang, P., Wee, C. W., Whyard, S., Zhang, W., Nagaraju, J., ffrench-Constant, R. H., Herrero, S., Gordon, K., Swevers, L. and Smagghe, G.

(2011). RNA interference in Lepidoptera: An overview of successful and unsuccessful studies and implications for experimental design. *Journal of Insect Physiology*, 57:231–245.

Terra, W. R., Ferreira, C. and Baker, J. E. (1996). Compartmentalization of digestion, In: Lehane, M. J. and Billingsley, P. F. (eds.), *Biology of the Insect Midgut.* Springer Netherlands, pp. 206–235.

Tian, G., Cheng, L., Qi, X., Ge, Z., Niu, C., Zhang, X. and Jin, S. (2015) Transgenic cotton plants expressing double-stranded RNAs target *HMG-CoA Reductase* (*HMGR*) gene inhibits the growth, development and survival of cotton bollworms. *International Journal of Biological Sciences*, 11(11): 1296-1305.

Tomoyasu, Y., Miller, S. C., Tomita, S., Schoppmeier, M., Grossmann, D. and Bucher G. (2008). Exploring systemic RNA interference in insects: a genome-wide survey for RNAi genes in *Tribolium. Genome Biology*, 17; 9(1):R10.

Trivedi, B. (2010). Bug silencing: the next generation of pesticides. *The New Scientist*, 205: 34–37.

Turner, C. T., Davy, M. W., Macdiarmid, R. M., Plummer, K. M., Birch, N. P. and Newcomb, R. D. (2006). RNA interference in the light brown apple moth, *Epiphyas postvittana* (Walker) induced by double-stranded RNA feeding. *Insect Molecular Biology*, 15:383–391.

Twari, M., Sharma, D., Trivedi, P. K. (2014) Artificial microRNA mediated gene silencing in plants: progress and perspectives. *Plant Molecular Biology*, 86:1-18.

Ulvila, J., Parikka, M., Kleino, A., Sormunen, R., Ezekowitz, R. A., Kocks, C. and Ramet, M. (2006). Double-stranded RNA is internalized by scavenger receptor-mediated endocytosis in Drosophila S2 cells. *Journal of Biological Chemistry*, 281:14370–14375.

Umesh Balkrishna Jagtap, U. B., R Gurav, R. G. and Vishwas Bap, V. A. (2011). Role of RNA interference in plant improvement. *Naturwissenschaften* 98:473–492.

Upadhyay, S. K., Chandrashekar, K., Thakur, N., Verma, P. C., Borgio, J. F., Singh, P. K. and Tuli, R. (2011). RNA interference for the control of whiteflies (*Bemisia tabaci*) by oral route. *Journal of Biosciences*, 36:153–161.

Wang, K., Peng, Y., Pu J., Fu., W, Wang, J. and Han, Z. (2016). Variation in RNAi efficacy among insect species is attributable to dsRNA degradation in vivo. *Insect Biochemistry and Molecular Biology*, 77:1-9.

Wang, X., Yang, Y., Zhou, J., Yu, C., Cheng, Y., Yan, C. and Chen, J. (2012). Two step method for constructing *Arabidopsis* artificial microRNA vectors. *Biotechnology Letters* 34:1343–1349.

Wang, Y. B., Zhang, H., Li, H. C. and Miao, X. X. (2011). Second generation sequencing supply an effective way to screen RNAi targets in large scale for potential application in pest insect control. *PLoS One*, 6(4): e18644.

Whangbo, J. S. and Hunter, C. P. (2008). Environmental RNA interference. *Trends in Genetics*, 24:297–305.

Xiao, Y. H., Yin, M. H., Hou, L. and Pei, Y. (2006). Direct amplification of intron-containing hairpin RNA construct from genomic DNA. *Biotechniques*, 41:548–552.

Xiong, Y. (2013). Silencing the HaHR3 gene by transgenic plant-mediated RNAi to disrupt *Helicoverpa armigera* development. *International Journal of Biological Sciences*, 9:370–381.

Xu, W. and Han, Z. (2008). Cloning and phylogenetic analysis of sid-1-like genes from aphids. *Journal of Insect Science*, 8:30.

Yan, H., Deng, X., Cao, Y., Huang, J., Ma, L. and Zhao, B. (2011). A novel approach for the construction of plant amiRNA expression vectors. *Journal of Biotechnology*, 151:9–14.

Yan, P., Shen, W., Gao, X., Li, X., Zhou, P. and Duan, J. (2012). High-throughput construction of intron-containing hairpin RNA vectors for RNAi in plants. *PLoS One* 7:e38186.

Yang, J., and Han, Z. J. (2014). Efficiency of different methods for dsRNA delivery in cotton bollworm (*Helicoverpa armigera*). *Journal of Integrative Agriculture*, 13 (1):115–123.

Yogindran, S. and Rajam M. V. (2016). Artificial miRNA-mediated silencing of ecdysone receptor (EcR) affects larval development and oogenesis in *Helicoverpa armigera. Insect Biochemistry and Molecular Biology*, 77: 21-30.

Zha, W., Peng, X., Chen, R., Du, B., Zhu, L. and He, G. (2011). Knockdown of midgut genes by dsRNA-transgenic plant-mediated RNA interference in the hemipteran insect *Nilaparvata lugens. PLoS One*, 6 (5): e20504.

Zhang, H., Li H. C. and Miao X. X. (2012). Feasibility, limitation and possible solutions of RNAi-based technology for insect pest control. *Insect Science*, 00:1-16.

Zhang, J., Khan, S. A., Hasse, C., Ruf, S., Heckel, D. G. and Bock, R. (2015). Full crop protection from an insect pest by expression of long double-stranded RNAs in plastids. *Science* 347: 991–994.

Zhang, L., Shang, Q., Lu, Y., Zhao, Q. and Gao, X. (2015). A transferrin gene associated with development and 2-tridecanone tolerance in *Helicoverpa armigera. Insect Molecular Biology*, 24:155–166.

Zhao, Y. Y., Liu, F., Yang, G. and You, M. S. (2011). PsOr1, a potential target for RNA interference-based pest management. *Insect Molecular Biology*, 20:97–104.

Zhou, J., Yu, F., Chen, B., Wang, X., Yang, Y., Cheng, Y., Yan, C. and Chen, J. (2013a) Universal vectors for constructing artificial microRNAs in plants. *Biotechnology Letters*, 35:1127–1133.

Zhou, J., Yu, F., Chen, B., Wang, X., Yang, Y., Cheng, Y., Yan, C. and Chen, J. (2013b). Two step method for constructing *Arabidopsis* artificial microRNA vectors. *Biotechnology Letters*, 34:1343–1349.

Zhu, J. Q., Liu, S., Ma, Y., Zhang, J. Q., Qi, H. S., Wei, Z. J., Yao, Q., Zhang, W. Q. and Li, S. (2012). Improvement of pest resistance in transgenic tobacco plants expressing dsRNA of an insect-associated gene EcR. *PLoS One*, 7: e38572.

In: Functional Genomics ISBN: 978-1-53612-565-8
Editors: Holger Uffe and Olaf Philip © 2017 Nova Science Publishers, Inc.

Chapter 2

TARGETED GENOME ENGINEERING BY CRISPR/CAS9: THE NEW PERSPECTIVE OF FUNCTIONAL GENOMICS

Era Vaidya Malhotra, Anshul Watts, Archana Watts, Nikhil Ram Kumar, Shweta Singh, Pragya Mishra, Ravi Prakash Saini, Kesiraju Karthik, Suraj Singh, R. Maniraj, N. Muralimohan, T. Arulprakash, Saurabh Tyagi, Debasis Pattanayak and Rohini Sreevathsa*

ICAR-National Research Centre on Plant Biotechnology,
Pusa Campus, New Delhi, India

ABSTRACT

Precise genome modification, in a predictable manner holds the key to bring in a new dawn in basic and applied research. Targeted genome editing using engineered nucleases is fast emerging as the foremost

---

* Corresponding author Email: rohinisreevathsa@gmail.com.

technique for elucidating and manipulating gene functions. Rapid advances in genome editing are being made after the emergence of the clustered, regularly interspaced, short palindromic repeats (CRISPR) – CRISPR-associated protein9 (Cas9) technology. This technique has been adapted from the microbial adaptive immune system, and is now being extensively used in developing different human therapeutics, alleviating genetic disorders, livestock breeding and developing improved high yielding plants. The system exploits the property of the RNA-guided endonuclease (Cas9) to cleave double stranded DNA in the targeted genomic loci under the direction of easily reprogrammable guide RNAs (gRNAs). Once the nicks are induced, they are either repaired by the cells' nonhomologous end-joining (NHEJ) repair mechanism, or sequence replacement by homology-directed repair (HDR). This technology can also be used to stimulate the function of a gene by fusing its activation domain to a catalytically inactive Cas9 and targeting the promoter region. This chapter shall in detail cover the origin of this technology, its adaptation from prokaryotic to eukaryotic systems, evolution and applications. The use of this system for functional genomics studies and applications of CRISPR-Cas9 in crop improvement will also be described. CRISPR-Cas9 has evolved to become the face of genetic engineering, and is set to revolutionize functional genomics in the near future.

Keywords: CRISPR/Cas, Cas9 nuclease, genome editing, TALEN, ZFNs

1. Introduction

Crop improvement is a continuous process and a large number of techniques are employed to keep the improvement process dynamic. Over the decades, conventional plant breeding has been used for accelerating crop improvement, improving agronomic performance of crops and accumulating favourable alleles in the crop gene pool. Traditional breeding strategies depend on mutagenesis to generate new variability into the existing gene pool. But this method is time consuming and requires screening of large populations. With the advancement of technologies and increase of scientific know how, high throughput techniques have emerged which are highly efficient in modifying genomes at precisely known locations.

The most recent addition to the genome modification techniques are the principles of genome editing. Sequence specific nucleases have made the editing of genomes of living cells possible. These techniques use engineered endonucleases which contain sequence-specific DNA-binding domains or RNA sequences that cleave DNA in a sequence-specific manner. The double stranded breaks generated in the DNA result in activation of cellular DNA repair mechanisms, leading to modifications at the target sites.

The genome editing techniques enable scientists to modify gene expression and regulation, thus facilitating functional genomics studies. Zinc finger nucleases (ZFN) and transcription activator-like effector nucleases (TALEN) comprise the earliest generations of targeted genome editing tools. The CRISPR/Cas9 technology has now emerged as the next promising tool for genome editing, and is derived from the bacterial type-II CRISPR/Cas adaptive immune system. The term CRISPR is an acronym for Clustered Regularly Interspaced Short Palindromic Repeats and Cas implies CRISPR-associated protein. CRISPR has the ability to speed up crop improvement programmes as it allows for the introduction of precise and predictable mutations directly in cultivated plant varieties.

The impact of this technology is wide spread and it is being ushered into the genome editing world with great enthusiasm. CRISPR/Cas9 is being utilised for the modification of economically significant plants because of its technical viability, accuracy and public acceptance.

2. CRISPR/Cas System: Discovery

CRISPR/Cas systems are an adaptive immune mechanism of many prokaryotes, used as a defence response against foreign entities such as bacteria, phages, viruses and plasmids (Piatek et al., 2014). It works by degrading foreign DNA in a sequence dependent manner. The CRISPR loci were discovered in *Escherichia coli* in 1980s (Ishino et al., 1987; Mojica et al., 2000), but their functions were confirmed much later. The acronym CRISPR was coined in 2002 after it was discovered that such arrays were present in several bacteria and archaea (Jansen et al., 2002a; Jansen et al., 2002b). They have been detected in 45% of sequenced bacterial genomes and 90% of sequenced archaea (Grissa et al., 2007; Horvath and Barrangou, 2010). Parallelly many different well conserved Cas genes have been found to be present adjacent to the repeat elements (Jansen et al., 2002a).

The year 2005 brought CRISPR into the limelight, with the reports of sequence homology of the CRISPR spacer sequences to DNA sequences of viruses and plasmids (Bolotin et al., 2005; Mojica et al., 2005; Pourcel et al., 2005). This led to the hypothesis that CRISPRs and the associated proteins might have some role in bacterial immunity. The confirmation was provided by observations of the involvement of CRISPR-Cas in providing virus resistance (Barrangou et al., 2007) and in preventing plasmid transfer (Marraffini and Sontheimer, 2008).

The CRISPR loci contain short tandem repeats of genomic DNA and unique protospacer sequences derived from sequences of the invading foreign DNA. These function in close association with the Cas proteins (Figure 1). When any foreign DNA invades the bacterial cell, the CRISPR system recognises the invader using the spacer sequences as a genetic memory and Cas9 then destroys the foreign entity. The number of spacers in a CRISPR array varies from one to several hundred. The repeats may vary from 21-48bp in length, and the spacers have a length between 26bp and 72bp. The Cas proteins have identifiable characteristic domains of helicases, nucleases, polymerases and RNA-binding proteins.

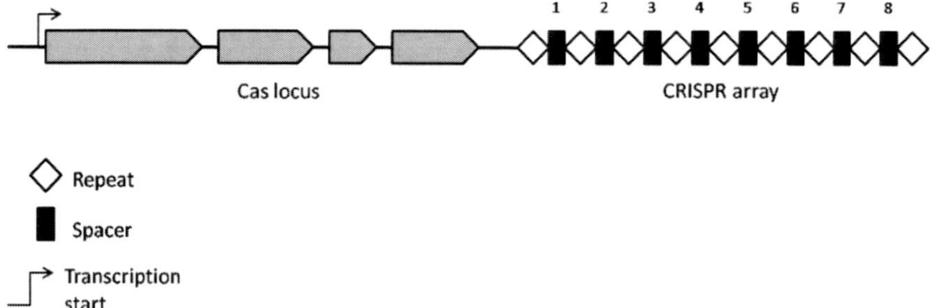

Figure 1. Structure of a CRISPR locus.

Three types of CRISPR mechanisms have been discovered, Type I, Type II and Type III, and the Type II mechanism discovered in *Streptococcus pyogenes* is the most widely studied type (Jinek et al., 2012; Bortesi and Fischer, 2015). Types I and III utilize specific Cas endonucleases for generating pre-crRNAs (Pre-CRISPR RNA), which then assemble into the Cas protein complex on maturity. This complex recognises and cleaves sequences complementary to the crRNA. In the Type II system, nucleic acid sequences from the invading DNA are incorporated within the CRISPR repeats in the host genome. On subsequent encounter with the same foreign entity, the sequences from the CRISPR arrays are transcribed as CRISPR RNA (crRNA) of approximately 40nt length. Each crRNA contains a transcribed sequence from the invading DNA, known as the 'protospacer,' incorporated into the CRISPR locus on previous invasion, and a part of the CRISPR repeat sequence. Another RNA, transactivating CRISPR RNA (tracrRNA) then hybridizes with each crRNA, followed by binding of this complex with the Cas9 nuclease (Barrangou et al., 2007). The protospacer part of the crRNA regulates Cas9 mediated cleavage of complementary target sequences upstream to short sequence stretches known as protospacer adjacent motifs (PAMs). Cas9 induces DNA double-strand breaks (DSBs) at the specific loci only (Reeks et al., 2013; Barrangou and Marraffini, 2014). A 'seed sequence' located 12 bases upstream of the PAM maintains the cleavage specificity. This sequence must match between the RNA and target DNA for Cas9 to function.

3. Mechanism of the CRISPR/Cas System

CRISPR-Cas system degrades the foreign DNA in a sequence-specific manner. The mechanism can be divided into three phases (Figure 2):

(i) adaptation
(ii) crRNA biogenesis
(iii) target interference

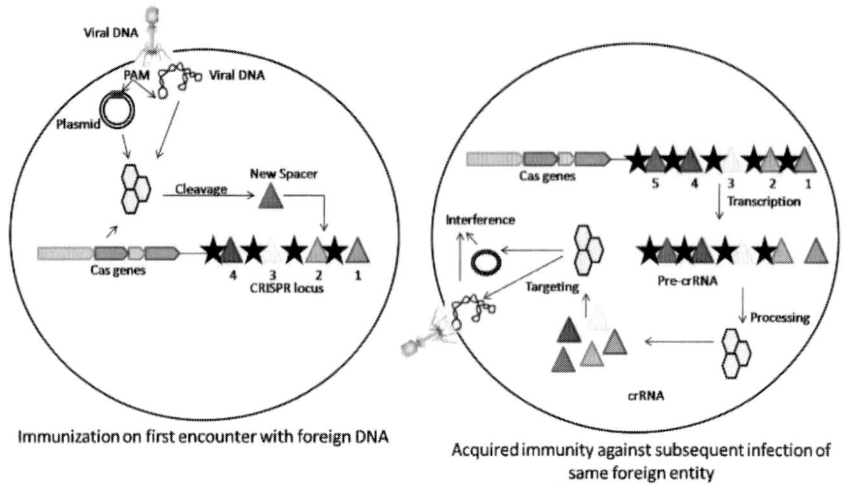

Figure 2. CRISPR/Cas mechanism.

3.1. Stage I – Adaptation

In the first stage, known as adaptation, immunization or spacer acquisition, a particular sequence of the invading foreign DNA is recognised and subsequently integrated as a spacer sequence, now referred to as the protospacer (Deveau et al., 2008) between two adjacent repeats in the CRISPR array. The spacers are usually integrated in a sequential manner at the leader end of the CRISPR array so the position of the spacer can infer the timeline of spacer acquisition. Spacer acquisition may be naive, i.e., when the invader is encountered for the first time, or primed, when there is a pre-existing spacer of the invader in the CRISPR due to previous infection. Two proteins, Cas1 and Cas2, are required in the acquisition step, and are present in all the cells having a CRISPR/Cas system. Both these nucleases form a complex for acquisition, but the catalytic function of Cas1 is indispensable for acquisition while that of Cas2 is not. Many other enzymes and proteins have found to be involved, one of which is Cas9 which plays a role in guiding the integration machinery. The target sequences have a short motif known as the Protospacer Adjacent Motif (PAM), adjacent to them, which are recognised by Cas9 in the Type II-A system. After protospacer identification, Cas9 regulates Cas1, Cas2 and Csn2 activity leading to incorporation of the new spacer into the CRISPR array. In the Type I-E system PAMs are recognised by Cas1-Cas2 complex.

3.2. Stage II – Biogenesis

The second stage is referred to as expression or crRNA biogenesis. The CRISPR array is transcribed into the precursor crRNA (pre-crRNA) by RNA polymerase which is further processed into small mature CRISPR RNAs (crRNAs) by specific endoribonucleases. The crRNAs contain a single spacer flanked by partial repeats. The Cas protein varies with the CRISPR system type. In cases of co-existence of the three CRISPR-Cas types, they do not interfere in processing of each other's pre-crRNA (Carte et al., 2014). The Cas6 family performs the processing step in type I and III systems. They generate crRNAs which are flanked by a short 5′ tag. In type I-C system, Cas5d is involved in generating crRNAs with an 11nt 5′ tag. Then the 3′ end of these crRNAs is processed and mature crRNAs contain a full spacer portion (5′ end) and a repeat-portion (3′ end) (Hale et al., 2009; Charpentier et al., 2015). In the type II systems a host RNase III, a trans-encoded small RNA (tracrRNA) that base pairs with the precrRNA and the Cas9 protein play major roles (Deltcheva et al., 2011; Jinek et al., 2012). The tracrRNA is complementary to the repeat sequence, and hence it forms an RNA duplex with each of the repeats of the pre-crRNA. The duplex is stabilized by Cas9, and is recognised and further processed by the host RNaseIII to produce intermediate forms of crRNA. The intermediate forms are further processed to yield the mature guide RNA (Deltcheva et al., 2011).

3.3. Stage III – Interference

In the last stage, known as target interference or immunity, the mature crRNAs are used to interfere with the invading foreign nucleic acids. The crRNA - Cas protein complex identifies the corresponding protospacer and triggers target cleavage by specific Cas nucleases. The type I system uses Cascade (CRISPR-associated complex for antiviral defence)-like complexes for target degradation, while the type II systems, have single effector protein for target interference. In type I systems the Cascade identifies the foreign DNA, Cas3 then induces a nick in the foreign DNA and subsequently degrades it (Westra et al., 2012). Both type I and type II systems necessarily require the PAM sequence, and perfect protospacer-crRNA complementarity in the seed region. Cascade interacts with the DNA and scans for PAMs and seed regions. Another protein, CseI, detects the desired PAM. After pairing of the seed region, the rest of the crRNA is base paired, displacing the non-bound DNA strand leading to formation of an R-loop. The changes in Cascade and target DNA due to crRNA binding trigger the Cas3 for target DNA cleavage. In type II systems, the Cas9 guided by the tracrRNA and crRNA introduces DSB in the target DNA (Jinek et al., 2012). Cas9 has separate target recognition and nuclease activity lobes. The target DNA-crRNA heteroduplex binds to the recognition lobe while the HNH and RuvC nuclease domains that cleave the complementary and non-complementary strands of the target, respectively are present on the nuclease lobe. The crRNA binding to Cas9 activates it by inducing conformational changes of the structural lobes, facilitating target DNA binding. The type III systems utilize the Cas10-Csm or Cas10-Cmr complexes. No PAMs are detected for Type III systems.

Hence, for the CRISPR/Cas system to function as a defence mechanism, all the three stages are required to be functional and interrelated. If the foreign DNA goes unrecognised, then the foreign entity proliferates and the host cell loses its immunity.

4. EVOLUTION OF CRISPR/CAS9 AS A GENOME EDITING TOOL

Genome editing has come to the forefront in the recent years. It includes a set of many diverse tools involving targetable endonucleases, applicable for targeted genome modification. These techniques facilitate changes in precise gene expression regulation at pre-determined sites. This technology revolves around two main elements: programmable nucleases that induce DNA DSB into desired sites and the cellular DSB repair mechanisms. Two repair pathways are operational in cells, the error prone non- homologous end joining (NHEJ) and the relatively precise homology-directed repair (HDR). Thus, using these features, desired mutations can be introduced into the genomic regions of interest. ZFNs were the first to be employed for editing genomes in 1996 (Kim et al., 1996), and TALENs were introduced in 2010 (Christian et al., 2010), followed by the application of CRISPR/Cas9 for genome editing in 2013 (Cong et al., 2013; Mali et al., 2013). Till date almost all CRISPR/Cas9 based technologies use the type II *Streptococcus pyogenes* derived system. Among the other techniques, the versatility, precision and specificity offered by CRISPR/Cas system has led to its evolution as the principal method of genome editing. The CRISPR based system does not use synthetic DNA-binding domains making it less complex in comparison to other methods of genome editing such as ZFNs and TALENS.

Initial reports on genome editing involved oligonucleotide mediated mutagenesis (OMM) and was used for introducing herbicide resistance in rice, maize, and tobacco (Kochevenko and Willmitzer, 2003; Iida and Terada, 2005). Engineered meganucleases (EMN) were then used in maize (Gao et al., 2010). Native endonucleases were tweaked to create very specific DSBs at defined loci, resulting in disruption of genes. After this, the ZFN and TALEN based approaches took the lead in targeted genome modification and yielded some good results. However due to the complexity in the design of ZFN and TALEN these technologies are relatively difficult to use in routine applications. Finally, in 2013

CRISPR/Cas9 system was for the first time successfully used for genome editing in eukaryotic cells. CRISPR/Cas9 was applied for the first time for growth and yield improvement in rice (Miao et al., 2013; Zhang et al., 2014). In 2016, the first CRISPR based herbicide tolerant canola, Cibus 5715, was approved for cultivation in Canada. Now several research communities are utilizing CRISPR/Cas9 in a wide array of crops for developing crops with improved traits.

A timeline of evolution of CRISPR/Cas9 for genome editing is traced in Figure 3.

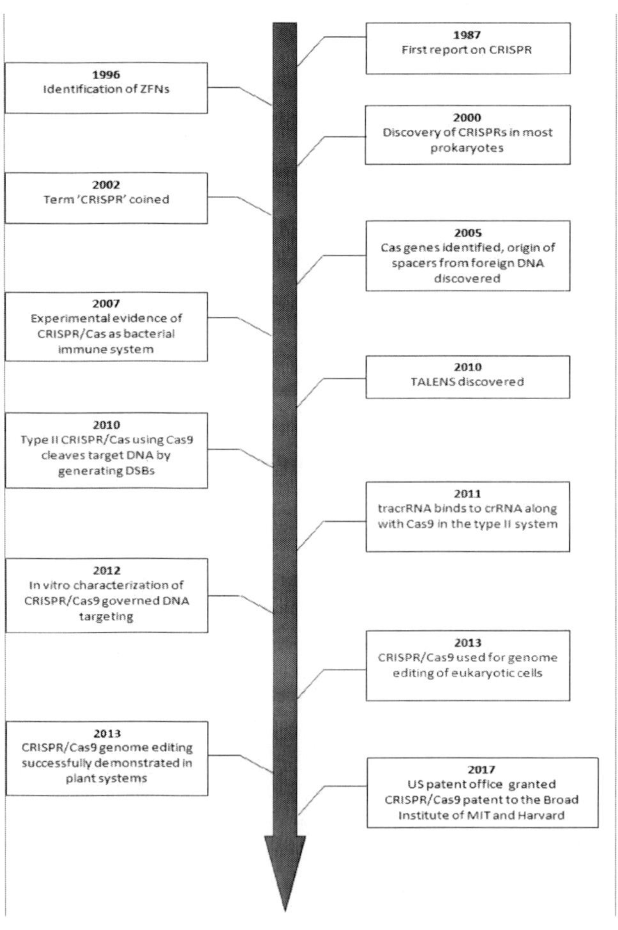

Figure 3. Timeline of development of genome editing by CRISPR/Cas9.

4.1. CRISPR/Cas as Genome Editing Tool

CRISPR/Cas system rose to become a genome editing wonder tool with the demonstration that any genomic sequence could be targeted by altering 20bp in the crRNA and a chimeric single guide RNA (gRNA) could be formed by combining the crRNA and tracrRNA. The gRNA consists of the 20bp DNA-recognition sequence at the 5′ end and a tracrRNA-derived region at the 3′ end. A tetraloop is formed by a four nucleotide (5′-GAAA-3′) linker between the crRNA region and the tracrRNA region. Additionally, there are minor loops on the 3′side of the tracrRNA region, useful for high gRNA expression. The DNA-recognition sequence in the gRNA, 3bp PAM and a 12bp seed sequence are responsible for target specificity of CRISPR/Cas9 system. For using CRISPR/Cas system for genome editing, there are two vital requirements. The first is the guide RNA (gRNA); and the second is a functional Cas9 protein. Cas9 requires a specific PAM sequence. The Cas9-gRNA complex scans for the PAM in the genome, and once located, it unwinds the DNA for initiating DNA-RNA base-pairing. Then the HNH and RuvC domains of Cas9, induce a nick in the DNA, each cleaving one strand, creating a linear DSB between the bases 3- and 4-bp upstream of the PAM sequence. The cellular repair machinery repairs the DSB, and gene mutagenesis or replacement can be obtained during this repair process via the NHEJ pathway or HDR pathway, depending on the availability of a DNA repair template.

The NHEJ pathway induces random insertions or deletions in the region where a DSB occurs, thus causing frameshift mutations when the targeted region lies in the coding region of a gene, leading to a gene knockout. On the other hand, if a template with sequence homology to the DSB surrounding sequence is available, the HDR pathway comes into play, leading to the possibility of targeted introduction of a desired sequence of interest in that particular genomic location.

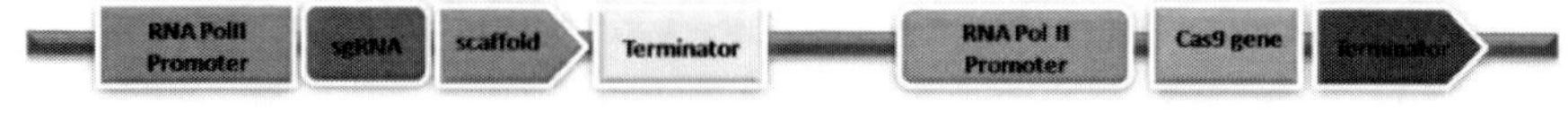

Figure 4. Typical features of CRISPR/Cas9 cassette.

4.2. Development of CRISPR/Cas9 Vector

Successful genome editing using CRISPR/Cas9 requires development of a vector system which contains all the components such as sgRNA specific for target genomic region, cas9 gene, promoters for expressing cas9 and sgRNA, terminator for sgRNA and sgRNA scaffold sequence. Cas9 cassette (Promoter +Cas9 gene +Terminator) and sgRNA cassette (Promoter+sgRNA+sgRNAscaffold+Terminator) can either be used in a separate vector or in a single vector (Figure 4). Several CRISPR/Cas9 vectors are available at Addgene (www.addgene.org, a non-profit plasmid repository).

4.2.1. gRNA

The first and foremost requirement of genome editing is the 20nt gRNA which should be complementary to the target genomic region. However, sufficient care must be taken while designing gRNA so that it is unique to the target region and should not show complementarity with any other genomic region. In the targeted genomic region, downstream to this 20 bp gRNA region, PAM region consisting of NGG or NAG should be present. This PAM sequence is not required in gRNA but it essential in the genomic DNA for recognition of target sequence by gRNA. For knockout of the target gene, gRNA should be preferably located in the 5′ region of the gene. gRNAs can be designed using various software tools, such as ECRISP (http://www.e-crisp.org/E-CRISP/) (Heigwer et al., 2014), CRISPR design tool (http://crispr.mit.edu/), CasOT (http://casot.cbi.pku.edu.cn/#casot) (Xiao et al., 2014) and CHOPCHOP (https://chopchop.rc.fas.harvard.edu/) (Montague et al., 2014). These softwares identify the region containing PAM sites, however they do not detect off targets. Apart from these, two softwares, i.e., CRISPR-P (http://cbi.hzau.edu.cn /cgi-bin/CRISPR2/CRISPR) (Lei et al., 2014) and CRISPR-Plant (http://www.genome.arizona.edu/crispr/CRISPRsearch.html) have been specifically designed to identify gRNA for plant genomes.

4.2.2. Promoter and Terminator

For achieving the optimal expression of both Cas9 and gRNA, they should be placed under a suitable promoter system. The gRNA is usually expressed from RNA polymerase III promoter while Cas9 is usually expressed from RNA polymerase II promoter. In dicots, usually CaMV 35S promoter is used to express the cas9 gene while in monocots both CaMV 35S and ubiquitin gene promoters are used. For expression of gRNA usually U3 or U6 promoters are used. U3 promoter has a preference for adenine (A) while U6 promoter has a preference for guanine (G) nucleotides. For sgRNA expression in dicots, U6 promoter is generally used. While in monocots, both U6 and U3 promoters are used. In a recent study in rice, CRISPR/Cas9 mediated genome editing efficiency was compared using U3 and U6 promoter and it was found that U6 promoter has more editing efficiency than U3 promoter (Milkami et al., 2015).

Apart from promoter, transcription termination sequences are also very important components of CRISPR/Cas9 vector. Cas9 gene is usually terminated by the nopaline synthase (NOS) gene terminator region (TNOS). While for gRNA a stretch of five to six thymine residues (TTTTTT) is used as the terminator sequence.

4.2.3. Cas9

Cas9 is an RNA guided DNA endonuclease that catalyses the site-specific cleavage of double stranded DNA. As the Cas9 gene is prokaryotic in origin, codon optimization is required for higher expression in eukaryotic systems. So, various modifications according to system of expression have been done in Cas9 gene. In the initial experiment, it was modified according to human codon biasness for expressing in human cells. Further, it was also modified according to monocot and dicot codon usage to express in monocots and dicots. In rice, using human codon optimized Cas9 an editing efficiency of around 2.5% was observed however, when plant codon optimized Cas9 was used the editing efficiency significantly increased up to 15% (Xu et al., 2014). So, codon optimized version of Cas9 is required to increase the editing efficiency.

Various kinds of modifications of Cas9 have been found for some special applications in molecular biology. The nuclease domain of Cas9 can be mutated. Such Cas9 in which nuclease domain is mutated and which is catalytically inactive and cannot cleave target region is known as dead cas9 (dCas9). Such dCas9 can also be used to regulate gene expression by co-expressing dCas9 and sgRNA which prevent transcription initiation or elongation by binding to the promoter or operator site of the gene so that RNA polymerase cannot bind. This strategy significantly decreases gene expression (Qi et al., 2013). dCas9 can be fused with activation or repression domain of a transcription factor thereby activating or repressing the transcription of a gene (Gilbert et al., 2014). dCas9 can also be fused with various epigenetic factors such as DNA methylated or histone modifying enzymes to regulate epigenetic control (Rusk, 2014). dCas9 can be fused to fluorescent proteins to check its localization on specific loci of the genome. Thus, it is a useful tool to study chromosome dynamics (Chen et al., 2013).

4.2.4. Targeting Multiple Genomic Regions through CRISPR/Cas9

One of the main advantage of CRISPR/Cas9 over other genome editing methods is the possibility of targeting multiple genomic loci or multiple genes simultaneously. For targeting multiple genes, single Cas9 is sufficient, however multiple sgRNA are required for each of the specific loci. One strategy used for this is to use separate RNA polymerase III promoters to express each sgRNA and separate each sgRNA from spacer sequence (Li et al., 2013) (Figure 5A). In the other strategy, each sgRNA is fused with a tRNA which is under the control of single RNA polymerase III promoter. This is transcribed as single transcript which is further processed by endogenous t-RNA processing RNases which would excise each sgRNA separately (Xie et al., 2015) (Figure 5B).

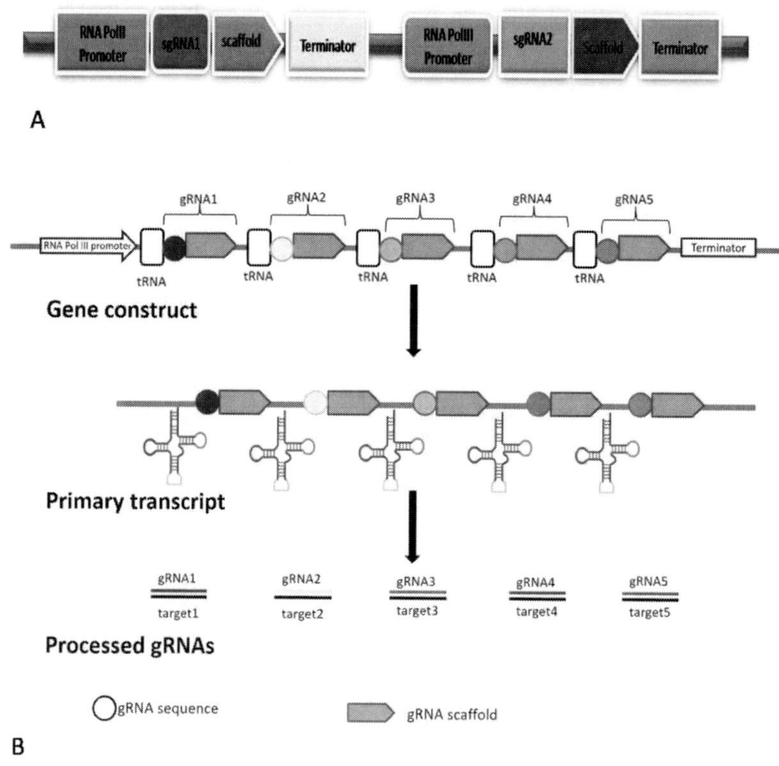

Figure 5. Multiple gene targeting through CRISPR/Cas9 system A) Each sgRNA has its own RNA polymerase III promoter B) Single RNA polymerase III promoter is used for all sgRNAs while each sgRNA is fused with tRNA.

4.3. Delivery of CRISPR/Cas9 Vector in Plant System

After the development of CRISPR/Cas9 vector, the next question is how to deliver it into plant cells. The most common methods used for delivery of CRISPR/Cas9 vector are *Agrobacterium* mediated transformation and particle gun bombardment methods. Apart from these, agroinfiltration, protoplast mediated and particle gun bombardment are also used for transient expression studies. Figure 6 highlights the protocol of CRISPR/Cas9 mediated genome editing of plant cells. After transformation, the resultant T_0 plants are screened for mutation either through PCR amplification and subsequent sequencing of that locus or through restriction digestion. Recently, a novel strategy of genome editing in plants has been shown in which instead of transforming with CRISPR/Cas9 vector, directly purified Cas9 protein and sgRNA is transformed into plant cells. This purified protein-RNA complex can induce genome editing without integrating into the plant cell. This strategy has additional advantage as the gene is not integrated in the host plant so the final product may not be considered as transgenic (Woo et al., 2015).

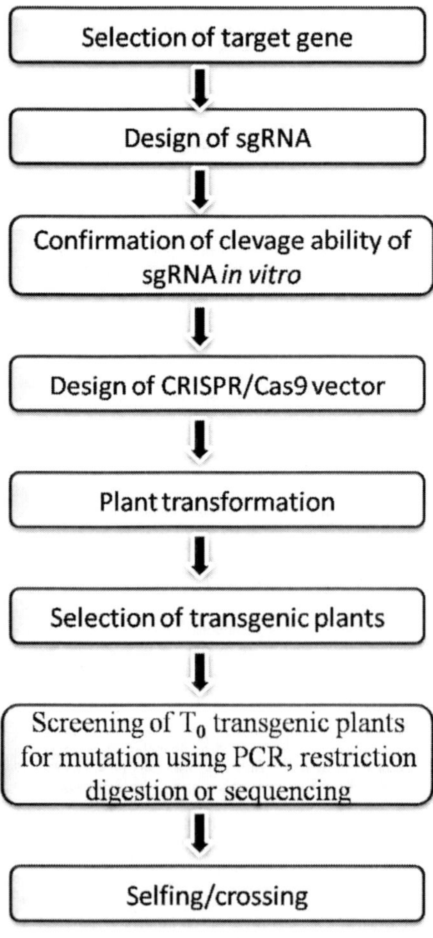

Figure 6. Protocol for CRISPR/Cas9 mediated genome editing of plant cell.

4.4. Applications of CRISPR/Cas9 in Functional Genomics

The applications of CRISPR/Cas9 are enormous and its scope is continuously expanding. The most common application of CRISPR/Cas9 is the generation of double stranded breaks in the target genomic region which is further repaired through NHEJ mechanism of the cell. This may cause insertions or deletions in the target genomic region. This strategy is commonly used as a gene knockout to functionally characterize a gene in reverse genetics. Further, if this NHEJ mediated heritable mutation occurs in the promoter or any other regulatory region, it can be exploited to understand the gene regulation and expression pattern. Through traditional functional genomics techniques like chemical or physical mutagenesis, transposon or T-DNA insertion, mutant lines can be generated. However, these are heterozygous mutant lines at T_0 stage which further require selfing for developing homozygous mutant lines. But through CRISPR/Cas9 homozygous mutant lines can be generated at T_0 stage. Further, other techniques generate random mutation but through CRISPR/Cas9 mutation is possible in a specific region of the genome.

Most of the agriculture crops are polyploid in nature and contain more than one copy of the same gene. Generating mutation in such crops for any gene is a difficult and cumbersome task. However, through CRISPR/Cas9 strategy homeologous genes can be easily mutated. In *B. napus* which is an allotetraploid, two *ALCATRAZ* (*ALC*) homeologs were simultaneously targeted through CRISPR/Cas9 using single gRNA (Braatz et al., 2017).

This technique is very useful in cytological studies like imaging, visualization of genomic loci and chromosome dynamics. For this nuclease-deactivated Cas9 (dCas9) has been fused with fluorescent reporter gene such as GFP (Ma et al., 2016). This technique can be employed to understand gene expression patterns by replacing the native promoter with another promoter (Liu et al., 2016). This technology is also very useful to study transcription activation, repression and RNA cleavage, gene function and regulation. Specific regions on the chromosome can be added or deleted, which is also a very important application in functional genomics. Disease causing genes or alleles can be replaced with beneficial genes by using this technology. Genes that are associated with specific traits can also be identified. Recently, in mammalian cells it has been shown that through this technology a single base can be mutated without inducing a double stranded break (Komor et al., 2016, Nishida et al., 2016). This technology can be used for large chromosomal deletion (100-200 kb). This becomes very useful in plant breeding when any unwanted loci can be deleted, which can result in increased plant performance. Additionally, this can also be used to study the function of gene clusters or large intergenic regions (Zhou et al., 2014).

4.5. Targeting CRISPR/Cas9 in Crop Plants

The first reports of using CRISPR/Cas9 for plant genome engineering came to the limelight in 2013 (Feng et al., 2013; Li et al., 2013; Nekrasov et al., 2013; Shan et al., 2013; Xie and Yang, 2013). Feng et al. (2013) demonstrated CRISPR/Cas9 system in the model plant *A. thaliana* and rice using transient and stable transgenic approach. In this study three Arabidopsis genes, namely, Brassinosteroid Insensitive 1 (BRI1), Jasmonate-zim-domain protein1 (JAZ1) and Gibberellic Acid Insensitive (GAI) and three rice genes, Outermost Cell-specific gene5 (ROC5), Stromal Processing Peptidase (SPP) and Young Seedling Albino (YSA), were knocked out using CRISPR/Cas9 system. Li et al. (2013) have targeted Arabidopsis and tobacco phytoene desaturase and flagellin sensitive2 gene using CRISPR/Cas9 system. Along with single gene, Li et al. (2013) have shown multiplexing by targeting two genes *AtRACK1b* and *AtRACK1c* simultaneously, using two sgRNA in a single cassette. Since then CRISPR/Cas9 system for genome editing has been modified by different research groups to suit their research interests and has emerged as the most widely used tool for generating crops having edited genomes for desired traits. An overview of publications reporting applications of the CRISPR/Cas9 system in plants is provided in Table 1.

Table 1. CRISPR/Cas9 mediated genome edited plants

Plant species	Target gene	Mutation frequency (%)	Reference
Arabidopsis thaliana	PDS3, FLS2	1.1–5.6	Li et al. (2013)
	PDS3	2.7	
	RACK1b, RACK1c	2.5-2.7	
	ADH1, TT4, RTEL1	26.7	Fauser et al. (2014)
	ADH1	42.8	Schiml et al. (2014)
	BRI1, GAI, JAZ1	30–84	Feng et al. (2013)
	YFFP	18.8	
	co-transfected GFP	58	Jia et al. (2014)
	Co-transfected GUUS, TT4, CHLI1, CHL12	38-89	Mao et al. (2013)
	Co-transfected GFP		Jiang et al. (2013a)
N.benthamiana	PDS3	4.8–38.5	Li et al. (2013)
	Co-transfected GFP		Jiang et al.(2013b)
	PDS	1.8–2.4	Nekrasov et al. (2013)
	PDS	12.7–13.8	Upadhyay et al. (2013)
N. tabacum	PDS, PDR6	16.27–20.3	Gao et al. (2014)
O. sativa	PDS, BADH2, MPK2, Os02g23823 OsPDS, OsBADH2	14.5–38.0 7.1–9.4	Shan et al. (2013)
	MPK5	3–8	Xie and Yang (2013)
	KO1, KOL5; CPS4, CYP99A2; CYP76M5, CYP76M6	-	Zhou et al. (2014)

Table 1. (Continued)

Plant species	Target gene	Mutation frequency (%)	Reference
	CAO1, LAZY1	83.3–91.6	Miao et al. (2013)
	PDS, PMS3, EPSPS, DERF1, MSH1, MYB5, MYB1, ROC5, SPP, YSA	21.1–66.7	Zhang et al.(2014)
	BAL	2 –16	Xu et al. (2014)
	OsU6 SWEET1a-1b-11-13	12.5–100	Zhou et al. (2014)
	SWEET14	-	Jiang et al.(2013b)
	FT	90	Hyun et al. (2015)
	Gnla, GS3, IPA1	42.5,57.5, 27.5	Li et al. (2016)
	ROC5, SPP, YSA	5–75	Feng et al. (2013)
	OsMYB1	50	Mao et al. (2013)
T. aestivum	MLO-A1	36	Wang et al. (2014)
	PDS, INOX	18–22	Upadhyayet al. (2013)
	MLO-A1	5.6	Brooks et al. (2014)
S. lycopersicum	GFP, SHR		Ron et al. (2014)
	SlAGO7, Solyc08g041770, Solyc07g021170, Solyc12g044760	75–100	Brooks et al. (2014)
	RIN	-	Ito et al. (2015)
	ATNI	29	Cermak et al. (2015)
	SlPDS	72-100	Changtian et al.(2016)
M. polymorpha	ARF1	11	Sugano et al. (2014)
Z. mays	LIG1	3.9	Svitashev et al.(2015)

Plant species	Target gene	Mutation frequency (%)	Reference
	Ms26 and Ms45	1	Svitashev et al.(2015)
	ALS1 and ALS2	1	
	PAT	1-2	
	IPK	16.4–19.1	Liang et al. (2014)
G. max	ALS1	59	Li et al. (2015)
	GmPDS	43.4-48.1	Du et al. (2016)
Populus tomentosa	PtoPDS	51.7	Fan et al. (2015)
	4CL1	100	Zhou et al. (2015)
	4CL2	100	
Petunia hybrid	PDS	55.6–87.5	Zhang et al. (2016)
Brassica oleracea	BolC.GA4.a	10	Lawrenson et al. (2015)
Hordeum vulgare	HvPM19	10-23	Lawrenson et al. (2015)
Solanum tuberosum	StIAA2	-	Wang et al. (2015)
Brassica napus	ALC	-	Braatz et al. (2017)

5. Comparison of CRISPR/Cas9 over ZFNs and TALENS

Before the advent and popularization of CRISPR/Cas9 for genome editing, ZFNs and TALENS were widely used for this purpose. Both of these require an individual protein with an engineered DNA-binding domain for every target, which is both laborious and expensive. On the other hand, the CRISPR/Cas system is simpler as it requires ribonucleotide complex formation, which can be done readily. Targeting efficiency, i.e., percentage of achieved desired mutation is much higher in CRISPR in comparison to both ZFNs and TALENS. ZFNs and TALENs depend on protein–DNA interactions for target specificity, while target sites in CRISPR/Cas9 are recognised by Watson–Crick base pairing.

In both, ZFNs and TALENS, the DNA binding domain of each polypeptide confers the sequence specificity, while the FokI nuclease domain carries out the cleavage. In contrast, sequence specificity in CRISPR is governed by the gRNA. ZFNs and TALENs contain the catalytic site of the restriction endonuclease FokI. The generated DSBs have overhangs with variable lengths. The nuclease domains of Cas9, i.e., RuvC and HNH, target the DNA nucleotides upstream of the PAM and result in blunt ends.

In addition, ZFNs have target site availability restrictions and lower target specificity. TALEN targets require a thymidine residue at the first position. However, the CRISPR/Cas9 system requires the PAM motif adjacent to the target sequence. Additionally, CRISPR/Cas9 has the ability to cleave the methylated DNA which other genome editing technique cannot do.

The main advantage offered by CRISPR/Cas9 is that of multiplexing. Many actions can be performed simultaneously, like introducing DSBs at many sites, editing of several genes at the same time, and knocking out of redundant genes or parallel pathways. For multiplexing CRISPR/Cas9 system requires only the Cas9 protein and different gRNAs. On the contrary, separate dimeric proteins are required for every target site for multiplexing with ZFNs or TALENs.

One of the main drawbacks of CRISPR/Cas9 is the off-target effects it causes. Initially, it was shown that 20 nt gRNA is required but later on it was shown that only 12-13 nt sequence within the sgRNA is required for target specific cleavage, rest mismatch can easily be tolerated. So, it can cause potential off-target effects also. For reducing off target effects some strategies have been designed. In one such strategy double nickase system is used in which two sgRNA are used one upstream to target sequence and one downstream to target sequence. Further, a modified Cas9 (cas9n, cas9 nickase) which is able to do single stranded break instead of double stranded break is used. This strategy shows significant reduction in off-target effect. This nicking strategy is similar to ZFN and TALEN where digestion takes place only when the two domains come together (Schiml et al., 2014, Mali et al., 2013). In another strategy, two sgRNA are used but instead of normal Cas9 an inactivated Cas9 is used which lacks nuclease activity. This Cas9 is fused with the nuclease domain of FokI and this strategy also reduced the off-target effects (Guilinger et al., 2014).

Overall, CRISPR/Cas9 system is advantageous over its predecessors due to its simplicity, accessibility, cost and versatility.

REFERENCES

Barrangou, R. & Marraffini, L. A. (2014). CRISPR-Cas systems: prokaryotes upgrade to adaptive immunity. *Molecular Cell*, 54: 234–244.

Barrangou, R., Fremaux, C., Deveau, H., Richards, M., Boyaval, P., Moineau, S., Romero, D. A. & Horvath, P. (2007). CRISPR provides acquired resistance against viruses in prokaryotes. *Science*, 315: 1709–1712.

Bolotin, A., Quinquis, B., Sorokin, A. & Ehrlich, S. D. (2005). Clustered regularly interspaced short palindrome repeats (CRISPRs) have spacers of extrachromosomal origin. *Microbiology*, 151:2551–2561.

Bortesi, L. & Fischer, R. (2015). The CRISPR system for plant genome editing and beyond. *Biotechnology Advances*, 33: 41–52.

Braatz, J., Harloff, H. J., Mascher, M., Stein, N., Himmelbach, A. & Jung, C. (2017). CRISPR-Cas9 targeted mutagenesis leads to simultaneous modification of different homoeologous gene copies in polyploid oilseed rape (*Brassica napus* L.) *Plant Physiology,* doi: http://dx.doi.org/10.1104/pp.17.00426.

Brooks, C., Nekrasov, V., Lippman, Z. B. & Van Eck, J. (2014). Efficient gene editing in tomato in the first generation using the clustered regularly interspaced short palindromic repeats/CRISPR-associated9 system. *Plant Physiology*, 166: 1292–1297.

Carte, J., Christopher, R. T., Smith, J. T., Olson, S., Barrangou, R., Moineau, S., Glover, C. V. 3rd, Graveley, B. R., Terns, R. M. & Terns, M. P. (2014). The three major types of CRISPR-Cas systems function independently in CRISPR RNA biogenesis in *Streptococcus thermophilus*. *Molecular Microbiology*, 93(1): 98–112.

Cermak, T., Baltes, N. J., Cegan, R., Zhang, Y. & Voytas, D.F. (2015). High-frequency, precise modification of the tomato genome. *Genome Biology*, 16: 232.

Changtian, P., Ye, L., Qin, L., Liu, X., He, Y., Wang, J., Chen, L. & Lu, G. (2016). CRISPR/Cas9-mediated efficient and heritable targeted mutagenesis in tomato plants in the first and later generations. *Scientific Reports*, 6: 24765.

Charpentier, E., Richter, H., van der Oost, J. & White, M.F. (2015). Biogenesis pathways of RNA guides in archaeal and bacterial CRISPR-Cas adaptive immunity. *FEMS Microbiology Reviews*, 39: 428–441.

Chen, B., Gilbert, L. A., Cimini, B. A., Schnitzbauer, J., Zhang, W., Li, G. W., Park, J., Blackburn, E. H., Weissman, J. S., Qi, L. S., and Huang, B. (2013). Dynamic imaging of genomic loci in living human cells by an optimized CRISPR/Cas system. *Cell*, 155: 1479–1491

Christian, M., Cermak, T., Doyle, E. L., Schmidt, C., Zhang, F., Hummel, A., Bogdanove, A. J. & Voytas, D. F. (2010). Targeting DNA double-strand breaks with TAL effector nucleases. *Genetics*, 186: 757–761.

Cong, L., Ran, F. A., Cox, D., Lin, S., Barretto, R., Habib, N., Hsu, P. D., Wu, X., Jiang, W., Marraffini, L. A. & Zhang, F. (2013). Multiplex genome engineering using CRISPR/Cas systems. *Science*, 339: 819–823.

Deltcheva, E., Chylinski, K., Sharma, C. M., Gonzales, K., Chao, Y., Pirzada, Z. A., Eckert, M. R., Vogel, J. & Charpentier, E. (2011). CRISPR RNA maturation by trans-encoded small RNA and host factor RNase III. *Nature*, 471(7340): 602–607.

Deveau, H., Barrangou, R., Garneau, J. E., Labonte, J., Fremaux, C., Boyaval, P., Romero, D. A., Horvath, P. & Moineau, S. (2008). Phage response to CRISPR encoded resistance in *Streptococcus thermophilus*. *Journal of Bacteriology*, 190: 1390–1400.

Du, H., Zeng, X., Zhao, M., Cui, X., Wang, Q., Yang, H., Cheng, H. & Yu, D. (2016). Efficient targeted mutagenesis in soybean by TALENs and CRISPR/Cas9. *Journal of Biotechnology*, 217: 90–97.

Fan, D., Liu, T., Li, C., Jiao, B., Li, S., Hou, Y. & Luo, K. (2015). Efficient CRISPR/Cas9-mediated targeted mutagenesis in Populus in the first generation. *Scientific Reports*, 5: 12217.

Fauser, F., Schiml, S. & Puchta, H. (2014). Both CRISPR/Cas-based nucleases and nickases can be used efficiently for genome engineering in *Arabidopsis thaliana*. *Plant Journal*, 79: 348–359.

Feng, Z., Zhang, B., Ding, W., Liu, X., Yang, D. L., Wei, P., Cao, F., Zhu, S., Zhang, F., Mao, Y. & Zhu, J. K. (2013). Efficient genome editing in plants using a CRISPR/Cas system. *Cell Research*, 23: 1229–1232.

Gao, H., Smith, J., Yang, M., Jones, S., Djukanovic, V., Nicholson, M. G., West, A., Bidney, D., Falco, S. C., Jantz, D. & Lyznik, L. A. (2010). Heritable targeted mutagenesis in maize using a designed endonuclease. *The Plant Journal*, 61: 176–187.

Gao, J., Wang, G., Ma, S., Xie, X., Wu, X., Zhang, X., Wu, Y., Zhao, P. & Xia, Q. (2014). CRISPR/Cas9-mediated targeted mutagenesis in *Nicotiana tabacum*. *Plant Molecular Biology*, 87(1): 99–110.

Gilbert, L. A., Horlbeck, M. A., Adamson, B., Villalta, J. E., Chen, Y. W., Whitehead, E. H., Guimaraes, C., Panning, B., Ploegh, H. L., Bassik, M. C., Qi, L. S., Kampmann, M. & Weissman, J. S. (2014). Genome-scale CRISPR-mediated control of gene repression and activation. *Cell,* 159: 1–15.

Grissa, I., Vergnaud, G. & Pourcel, C. (2007). CRISPRFinder: a web tool to identify clustered regularly interspaced short palindromic repeats. *Nucleic Acids Research*, 35 (Web Server issue): W52–W57.

Guilinger, J. P., Thompson, D. B. & Liu, D. R. (2014). Fusion of catalytically inactive Cas9to FokI nuclease improves the specificity of genome modification. *Nature Biotechnology*, 32: 577–582.

Hale, C. R., Zhao, P., Olson, S., Duff, M. O., Graveley, B. R., Wells, L., Terns, R. M. & Terns, M. P. (2009). RNA-guided RNA cleavage by a CRISPR RNA-Cas protein complex. *Cell*, 139: 945–956.

Heigwer, F., Kerr, G. & Boutros, M. (2014). E-CRISP fast CRISPR target site identification. *Nature Methods*, 11:122-123. (http://www.e-crisp.org/E-CRISP/).

Horvath, P. & Barrangou, R. (2010). CRISPR/Cas, the immune system of bacteria and archaea. *Science,* 327: 167–170.

Hyun, Y., Kim, J., Cho, S. W., Choi, Y., Kim, J. & Coupland, G. (2015). Site-directed mutagenesis in *Arabidopsis thaliana* using dividing tissue-targeted RGEN of the CRISPR/Cas system to generate heritable null alleles. *Planta*, 241: 271–284.

Iida, S. &Terada, R. (2005). Modification of endogenous natural genes by gene targeting in rice and other higher plants. *Plant Molecular Biology*, 59: 205–219.

Ishino, Y., Shinagawa, H., Makino, K., Amemura, M. & Nakata, A. (1987). Nucleotide sequence of the *iap* gene, responsible for alkaline phosphatase isozyme conversion in *Escherichia coli* and identification of the gene product. *Journal of Bacteriology*, 169: 5429–5433.

Ito, Y., Yokoi, A. N., Endo, M., Mikami, M. & Toki, S. (2015). CRISPR/Cas9-mediated mutagenesis of the RIN locus that regulates tomato fruit ripening. *Biochemical and Biophysical Research Communications*, 467: 76–82.

Jansen, R., Embden, J. D., Gaastra, W. & Schouls, L. M. (2002a). Identification of genes that are associated with DNA repeats in prokaryotes. *Molecular Microbiology*, 43: 1565–1575.

Jansen, R., van Embden, J. D., Gaastra, W. & Schouls, L. M. (2002b). Identification of a novel family of sequence repeats among prokaryotes. *OMICS*, 6: 23–33.

Jia, W., Yang, B. & Weeks, D. P. (2014). Efficient CRISPR/Cas9-mediated gene editing in *Arabidopsis thaliana* and inheritance of modified genes in the T2 and T3 generations. *PLoS One* 9: e99225.

Jiang, W, Bikard, D., Cox, D., Zhang, F. & Marraffini, L. A. (2013a). RNA-guided editing of bacterial genomes using CRISPR-Cas systems. *Nature Biotechnology*, 31: 233–239.

Jiang, W., Zhou, H., Bi, H., Fromm, M., Yang, B. & Weeks, D. P. (2013b). Demonstration of CRISPR/Cas9/ sgRNA-mediated targeted gene modification in Arabidopsis, tobacco, sorghum and rice. *Nucleic Acids Research*, 41: e188.

Jinek, M., Chylinski, K., Fonfara, I., Hauer, M., Doudna, J. A. & Charpentier, E. (2012). A programmable dual-RNA-guided DNA endonuclease in adaptive bacterial immunity. *Science*, 337: 816–821.

Kim, Y. G., Cha, J. & Chandrasegaran, S. (1996). Hybrid restriction enzymes: zinc finger fusions to Fok I cleavage domain. *Proceedings of the National Academy of Sciences USA*, 93: 1156–1160.

Kochevenko, A. & Willmitzer, L. (2003). Chimeric RNA/DNA oligonucleotide based site-specific modification of the tobacco acetolactate syntase gene. *Plant Physiology*, 132: 174–184.

Komor, A. C., Kim, Y. B., Packer, M. S., Zuris, J. A. & Liu, D. R. (2016). Programmable editing of a target base in genomic DNA without double stranded DNA cleavage. *Nature*, 533: 420–424.

Lawrenson, T., Shorinola, O., Stacey, N., Li, C., Ostergaard, L., Patron, N., Uauy, C., & Harwood, W. (2015). Induction of targeted, heritable mutations in barley and *Brassica oleracea* using RNA-guided Cas9 nuclease. *Genome Biology*, 16: 258.

Lei, Y., Lu, L., Liu, H.-Y., Li, S., Xing, F. & Chen, L. L. (2014). CRISPR-P: a web tool for synthetic single-guide RNA design of CRISPR-system in plants. *Molecular Plant*, 7: 1494-1496. (http://cbi.hzau.edu.cn/cgi-bin/CRISPR2/CRISPR).

Li, J. F., Norville, J. E., Aach, J., McCormack, M., Zhang, D., Bush, J., Church, G. M. & Sheen, J. (2013). Multiplex and homologous recombination-mediated genome editing in Arabidopsis and *Nicotiana benthamiana* using guide RNA and Cas9. *Nature Biotechnology*, 31: 688–691.

Li, M., Li, X., Zhou, Z., Wu, P., Fang, M., Pan, X., Lin, Q., Luo, W., Wu, G. & Li, H. (2016). Reassessment of the four yield related genes Gn1a, DEP1, GS3, and IPA1 in rice using a CRISPR/Cas9 system. *Frontiers in Plant Science*, 7: 377.

Li, Z., Liu, Z. B., Xing, A., Moon, B. P., Koellhoffer, J. P., Huang, L., Ward, R. T., Clifton, E., Falco, S. C. & Cigan, A. M. (2015). Cas9-Guide RNA directed genome-editing in soybean. *Plant Physiology*, 169: 960–970.

Liang, Z., Zhang, K., Chen, K. & Gao, C. (2014). Targeted mutagenesis in *Zea mays* using TALENs and the CRISPR/Cas system. *Journal of Genetics and Genomics*, 41: 63–68.

Liu, D., Hu, R., Palla, K. J., Tuskan, G. A. & Yang, X. (2016). Advances and perspectives on the use of CRISPR/Cas9 systems in plant genomics research. *Current Opinion in Plant Biology*, 30: 70–77.

Ma, H., Tu, L. C., Naseri, A., Huisman, M., Zhang, S., Grunwald, D. & Pederson, T. (2016). Multiplexed labeling of genomic loci with dCas9 and engineered sgRNAs using CRISPRainbow. *Nature Biotechnology*, 34: 528–530.

Mali, P., Aach, J., Stranges, P. B., Esvelt, K. M., Moosburner, M., Kosuri, S., Yang, L., & Church, G. M. (2013). CAS9 transcriptional activators for target specificity screening and paired nickases for cooperative genome engineering. *Nature Biotechnology*, 31: 833–838.

Mali, P., Yang, L., Esvelt, K. M., Aach, J., Guell, M., DiCarlo, J. E., Norville, J. E. & Church, G. M. (2013). RNA-guided human genome engineering via Cas9. *Science*, 339: 823–826.

Mao, Y., Zhang, H., Xu, N., Zhang, B., Gou, F. & Zhu, J. K. (2013). Application of the CRISPR-Cas system for efficient genome engineering in plants. *Molecular Plant*, 6: 2008–2011.

Marraffini, L. A. & Sontheimer, E. J. 2008. CRISPR interference limits horizontal gene transfer in staphylococci by targeting DNA. *Science*, 322: 1843–1845.

Miao, J., Guo, D., Zhang, J., Huang, Q., Qin, G., Zhang, X., Wan, J., Gu, H. & Qu, L. J. (2013). Targeted mutagenesis in rice using CRISPR-Cas system. *Cell Research*, 23: 1233–1236.

Mikami, M., Toki, S. & Endo, M. (2015). Comparison of CRISPR/Cas9 expression constructs for efficient targeted mutagenesis in rice. *Plant Molecular Biology*, 88: 561–572.

Mojica, F. J., Diez-Villasenor, C., Garcia-Martinez, J. & Soria, E. (2005). Intervening sequences of regularly spaced prokaryotic repeats derive from foreign genetic elements. *Journal of Molecular Evolution*, 60: 174–182.

Mojica, F. J., Díez-Villaseñor, C., Soria, E. & Juez, G. (2000). Biological significance of a family of regularly spaced repeats in the genomes of archaea, bacteria and mitochondria. *Molecular Microbiology*, 36: 244–246.

Montague, T. G., Cruz, J. M., Gagnon, J. A., Church, G. M. & Valen, E. (2014). CHOPCHOP a CRISPR/Cas9 and TALEN web tool for genome editing. *Nucleic Acids Research,* 42: W401–W407. (https://chopchop.rc.fas.harvard.edu/).

Nekrasov, V., Staskawicz, B., Weigel, D., Jones, J. D. & Kamoun, S. (2013). Targeted mutagenesis in the model plant *Nicotiana benthamiana* using Cas9 RNA-guided endonuclease. *Nature Biotechnology*, 31:691–693.

Nishida, K., Arazoe, T., Yachie, N., Banno, S., Kakimoto, M., Tabata, M., Mochizuki, M., Miyabe, A., Araki, M., Hara, K. Y., Shimatani, Z. & Kondo A. (2016). Targeted nucleotide editing using hybrid prokaryotic and vertebrate adaptive immune systems. *Science*, 353(6305) doi: 10.1126/science.aaf8729.

Piatek, A., Ali, Z., Baazim, H., Li, L., Abulfaraj, A., Al-Shareef, S., Aouida, M. & Mahfouz, M. M. (2014). RNA-guided transcriptional regulation *in planta* via synthetic dCas9-based transcription factors. *Plant Biotechnology Journal*, 13(4): 578–589.

Pourcel, C., Salvignol, G. &Vergnaud, G. (2005). CRISPR elements in *Yersinia pestis* acquire new repeats by preferential uptake of bacteriophage DNA, and provide additional tools for evolutionary studies. *Microbiology*, 151: 653–663.

Qi, L. S., Larson, M. H., Gilbert, L. A., Doudna, J. A., Weismann, J. S., Arkin, A. P. & Lim, W. A. (2013). Repurposing CRISPR as an RNA-guided platform for sequence-specific control of gene expression. *Cell*, 152: 1173–1183.

Reeks, J., Naismith, J. H. & White, M. F. (2013). CRISPR interference: a structural perspective. *Biochemical Journal*, 453: 155–166.

Ron, M., Kajala, K., Pauluzzi, G., Wang, D., Reynoso, M. A., Zumstein, K., Garcha, J., Winte, S., Masson, H., Inagaki, S., Federici, F., Sinha, N., Deal, R. B., Bailey-Serres, J. & Brady, S. M. (2014). Hairy root transformation using *Agrobacterium rhizogenes* as a tool for exploring cell type-specific gene expression and function using tomato as a model. *Plant Physiology*, 166: 455–469.

Rusk, N. (2014). CRISPRs and epigenome editing. *Nature Methods*, 11: 28.

Schiml, S., Fauser, F. & Puchta, H. (2014). The CRISPR/Cas system can be used as nuclease for in planta gene targeting and as paired nickases for directed mutagenesis in Arabidopsis resulting in heritable progeny. *Plant Journal*, 80(6): 1139–1150.

Shan, Q., Wang, Y., Li, J., Zhang, Y., Chen, K., Liang, Z., Zhang, K., Liu, J., Xi, J. J., Qiu, J. L. & Gao, C. (2013). Targeted genome modification of crop plants using a CRISPR-Cas system. *Nature Biotechnology*, 31: 686–688.

Sugano, S. S., Shirakawa, M., Takagi, J., Matsuda, Y., Shimada, T., Hara-Nishimura, I. & Kohchi, T. (2014). CRISPR/Cas9-mediated targeted mutagenesis in the liverwort *Marchantia polymorpha* L. *Plant Cell Physiology*, 55: 475–481.

Svitashev, S., Young, J. K., Schwartz, C., Gao, H., Falco, S. C. & Cigan, A. M. (2015). Targeted mutagenesis, precise gene editing, and site-specific gene insertion in maize using Cas9 and guide RNA. *Plant Physiology*, 169: 931–945.

Upadhyay, S. K., Kumar, J., Alok, A. & Tuli, R. (2013). RNA-guided genome editing for target gene mutations in wheat. *G3 (Bethesda)*, 3: 2233–2238.

Wang, S., Zhang, S., Wang, W., Xiong, X., Meng, F. & Cui, X. (2015). Efficient targeted mutagenesis in potato by the CRISPR/Cas9system. *Plant Cell Report*, 34: 1473–1476.

Wang, Y., Cheng, X., Shan, Q., Zhang, Y., Liu, J., Gao, C. & Qiu, J. L. (2014). Simultaneous editing of three homoeoalleles in hexaploid bread wheat confers heritable resistance to powdery mildew. *Nature Biotechnology*, 32: 947–951.

Westra, E. R., van Erp, P. B., Künne, T., Wong, S. P., Staals, R. H., Seegers, C. L., Bollen, S., Jore, M. M., Semenova, E., Severinov, K., de Vos, W. M., Dame, R. T., de Vries, R., Brouns, S. J. & van der Oost, J. (2012). CRISPR immunity relies on the consecutive binding and degradation of negatively supercoiled invader DNA by Cascade and Cas3. *Molecular Cell*, 46: 595–605.

Woo, J. W., Kim, J., Kwon, S. I., Corvalan, C., Cho, S. W., Kim, H., Kim, S. G., Kim, S. T., Choe, S. & Kim, J. S. (2015). DNA-free genome editing in plants with preassembled CRISPR-Cas9 ribonucleoproteins. *Nature Biotechnology*, 33: 1162–1164.

Xiao, A., Cheng, Z. C., Kong, L., Zhu, Z. Y., Lin, S., Gao, G. & Zhang, B. (2014). CasOT: a genome-wide Cas9/gRNA off-target searching tool. *Bioinformatics*, 30: 1180-1182.

Xie, K. & Yang, Y. (2013). RNA-guided genome editing in plants using a CRISPR-Cas system. *Molecular Plant*, 6: 1975–1983.

Xie, K., Minkenberg, B. & Yang, Y. (2015). Boosting CRISPR/Cas9 multiplex editing capability with the endogenous tRNA-processing system. *Proceedings of the National Academy of Sciences*, 112: 3570–3575.

Xu, R., Li, H., Qin, R., Wang, L., Li, L., Wei, P. & Yang, J. (2014). Gene targeting using the *Agrobacterium tumefaciens*-mediated CRISPR-Cas system in rice. *Rice*, 7: 5.

Zhang, B., Yang, X., Yang, C., Li, M. & Guo, Y. (2016). Exploiting the CRISPR/Cas9 system for targeted genome mutagenesis in petunia. *Scientific Reports*, 6: 20315.

Zhang, H., Zhang, J., Wei, P., Zhang, B., Gou, F., Feng, Z., Mao, Y., Yang, L., Zhang, H., Xu, N. & Zhu, J. K. (2014). The CRISPR/Cas9 system produces specific and homozygous targeted gene editing in rice in one generation. *Plant Biotechnology Journal*, 12: 797–807.

Zhou, H., Liu, B., Weeks, D. P., Spalding, M. H. & Yang, B. (2014). Large chromosomal deletions and heritable small genetic changes induced by CRISPR/Cas9 in rice. *Nucleic Acids Research*, 42: 10903–10914.

Zhou, X., Jacobs, T. B., Xue, L. J., Harding, S. A. & Tsai, C. J. (2015). Exploiting SNPs for biallelic CRISPR mutations in the outcrossing woody perennial Populus reveals 4-coumarate: CoA ligase specificity and redundancy. *New Phytologist*, 208: 298–301.

Web Sources

- www.addgene.org
- http://crispr.mit.edu

In: Functional Genomics ISBN: 978-1-53612-565-8
Editors: Holger Uffe and Olaf Philip © 2017 Nova Science Publishers, Inc.

Chapter 3

STRATEGIES FOR INSECT PEST MANAGEMENT IN PLANTS

***Nikhil Ram Kumar*[1], *Shweta Singh*[1], *Kesiraju Karthik*[1], *Pragya Mishra*[1], *Ravi Prakash Saini*[1], *Suraj Singh*[1], *R. Maniraj*[1], *N. Muralimohan*[1], *T. Arulprakash*[1], *Saurabh Tyagi*[1], *Era Vaidya Malhotra*[1], *T. Vinutha*[2], *Debasis Pattanayak*[1] and *Rohini Sreevathsa*[1,*]**

[1]ICAR- National Research Centre on Plant Biotechnology, New Delhi, India

[2]Division of Biochemistry, ICAR-Indian Agricultural Research Institute, New Delhi, India

ABSTRACT

Modern agriculture is already facing challenges of feeding the increasing global population, and many more difficulties lie ahead in the future. Mitigation of insects and pests is a herculean task so as to sustain productivity. This requires better pest management strategies, which

* Corresponding author e-mail: rohinisreevathsa@gmail.com.

would have a huge impact on the final productivity. This chapter discusses pest management strategies practiced over the past, such as agronomical approaches, modern breeding approaches, and use of chemical pesticides; for better defense strategies against key insect pests responsible for large loss of economics in terms of crop production. Various approaches based on genetic engineering including transgenics, and RNAi for insect resistance in plants have been described and followed. Utility of different biotechnological tools in the field of insect pest management will be discussed in this chapter, considering both benefits as well as limitations. Additionally, this chapter also puts forward an insight into recently developed novel approaches of genome editing for high throughput pest management.

Keywords: *Helicoverpa armigera*, CRISPR/Cas9, *Bacillus thuringenesis*, pesticides, wild relatives

1. Introduction

Agriculture in the 21st century faces a myriad of challenges in many agriculture-dependent developing countries, which emphasize on the production of more food and fibre to feed their population. Current trends suggest that the average daily energy requirement will be 3050 kcal per person by 2050 (2970 kcal in the third world countries), up from 2770 kcal in 2003/05. However, predictions suggest that production increase alone would not be sufficient to ensure food security for everyone, as our agricultural produce suffers a huge loss due to both abiotic and biotic stresses to crops.

Agricultural pests are defined as any animal or plant causing harm or damage to humans, domestic animals or crops, even if it only causes annoyance. Pests comprise a broad spectrum of organisms including insects, mites, ticks, mice, rats, and other rodents, nematodes, cestodes, weeds, fungi, bacteria, viruses (and other pathogens). These pests impact overall yield via different ways; by direct feeding, laying eggs underneath leaf tissue, etc. Their mode of action and the extent of damage also differ based on their need. As per their mouth parts, they chew, bite, scrape the top surface or simply suck the sap inside.

During their 350 million year co-evolution with plants, insects have evolved a variety of different interactions with plants (Gatehouse, 2002). These include insect-mediated pollination and seed dispersal that is beneficial to plants while others such as insect damage on plants due to feeding and spread of various deleterious viruses. Insect herbivores are mainly of two types: chewing type and sap-sucking type. Chewing type insects, e.g., *Helicoverpa armigera* (the cotton bollworm), *Spodoptera litura* (the tobacco caterpillar), etc. are considered as a major threat to plants. Sap sucking insects, although considered as secondary herbivores, are also deleterious to plants, e.g., aphids (*Aphis gossypiella*), white flies (*Bemissia tabacci*), etc. Chewing insects mainly damage plants by eating their parts whereas sap-sucking insects suck the phloem sap, spreading viruses. White flies alone spread more than 60 types of viruses (Fordyce, 2001).

Modern agricultural needs of higher yield and production have become the prime focus of researchers. This idea has sidelined the other factors important for the survival of plants, e.g., pest resistance traits. In competition with getting better productive cultivars through breeding processes, their other important characteristics have been sacrificed. In the present scenario, it is quite evident that many of the elite cultivars lack natural resistance to pests in comparison to their wild relatives. The negative effects of pest damage have encouraged an increasing reliance on chemical pesticides, but this has tended to drive a cycle in which each new pesticide keeps pace with the appearance of resistance to the previous pesticide in the insect population. Against this background, the possibility of developing insect resistant crops that could help reduce insect damage, while reducing the reliance on chemical insecticides, has a number of attractions.

In this chapter, we discuss various scientific strategies for achieving pest resistance along with the pros and cons related to the use of those strategies.

These insect resistance strategies are divided into three groups:

1. Strategies used from the ancient past:

Conventional and chemical strategies:
   a. Agronomical cultural practices
   b. Use of Pesticides
   c. Breeding approaches

2. Strategies used from the recent past like:
   a. Transgene technology
   b. RNAi technology

3. Future prospects of insect pest management:
   a. Genome Editing
   b. Wild relatives of crops

The focus of the chapter relies on all the above mentioned strategies.

2. Agronomical Cultural Practices

Agronomical cultural practices are the oldest methods that have been used to manage pest populations. These approaches are environment supportive and involve skill intensive techniques, such as the optimal design and management of agro-ecosystems in time and space (e.g., the design of integrated polyculture, management of environments, crop rotations, timing of sowing, harvesting and field operations, etc.), and the more heavy-handed interventions that can result in environmental degradation.

Cultural control practices make the environment less attractive and favourable for pest survival, spread, growth and reproduction. The objective is the achievement of reduction in pest count below economic injury levels.

Strategies or cultural practices were developed by focusing on the following objectives:

a. Artificially intervening in surrounding environment of crop to make it unfit for ovipositon by insects.
b. Altering timing of sowing of host crop to make it unavailable for pest.
c. Reduction in pest survival on the crop by encouraging its natural enemies, or by altering the crop's susceptibility to the pest.

Advantages: Agronomical cultural practices are normally the cheapest of all controlling measures because they usually only require modifications of normal production practices. Sometimes they only require organised and careful planning. These practices are the only control practices that are profitable for low value crops under large areas of cultivation. The success of agronomical practices is that they do not cause any detrimental side effects of pesticides, e.g., the development of resistance to pesticides, presence of undesirable residues in food and the environment, without killing of non-target organisms.

Disadvantages: Agronomical cultural practices require long-term planning for greatest effectiveness and they deserve careful timing. These practices rely on the substitution of knowledge and skills. These may be effective against one pest but may be ineffective against a closely related species of the same order. The effectiveness of cultural controls is difficult to assess and they do not always provide complete economic control of pests. Some cultural controls are economically not viable because low income farmer can't alter sowing timing. Since they are dependent on rain as source of irrigation, it is difficult for them to make any major changes in cultural practices.

Concerns over agronomical practices: Although, agronomical cultural practices are the oldest methods, with the advent of synthetic pesticides these strategies were rapidly de-emphasized and research on them was mostly discontinued. These cultural controls are preventative rather than curative. They are dependent on long-range planning which requires detailed knowledge of the bio-ecology of the crop, pests, natural controls and environment relations which were poorly understood. The results were not very variable, and it was often difficult to evaluate their effectiveness.

3. Pesticides

According to the Food and Agricultural Organization (FAO), any substance or mixture of substances aimed for preventing, destroying, or controlling any pest, species of plants or animals, causing harm during or otherwise interfering with the production, processing, storage of agricultural commodities are called as pesticides. Pesticides were classified in several ways; based on target organism (e.g., herbicides, insecticides, fungicides, rodenticides etc.), chemical structure (e.g., organic, inorganic, synthetic, or biological pesticide), and physical state (e.g., gaseous fumigant).

Advantages: Application of pesticides is the quickest remedy available for controlling crop pests. It is specific to the target pest and doesn't require any change in cultural practices for its uses. Pesticide application results in higher mortality of pests.

Disadvantages: Mostly synthetic pesticides leave very harmful impact on environment and its ecological balance. Majority of synthetic pesticides are non-biodegradable and persist in environment for longer period of time. Often, overdoses of chemical based pesticides kill non targeted pest e.g., pollinators. Apart from this, they cause the problem of biological magnification due entry of pesticide residues into the food chain.

Concerns over the use of pesticides: Considering the outcome of pesticide application thrust is now being put on opting for more environment friendly and sustainable approaches. Even though pesticides ensure high pest mortality but it is achieved at the cost of detrimental impact to environment.

Thus, researchers are shifting their focus to more eco-friendly methods for controlling pests.

4. Breeding for Pest Resistance

Most widely used method of pest control is the use of chemicals. However, despite its importance in the control of pests, the extensive use of chemicals is harmful for environment and has resulted in problems such as the selection of pest-insect populations that are resistant to the particular pesticide used, outbreaks of secondary pests, reductions in the population of natural enemies and pollinators, undesirable residues in foodstuffs, the poisoning in human beings, animals, and environmental contamination (Smith, 1970).

Integrated pest management might represent an option for minimizing insect attacks and maintaining the pest populations below the level of economic damage. One of the methods used for this purpose is the development of pest-resistant plant cultivars (Gallo et al., 2002). This method stabilizes productivity because it exhibits significant advantages over the use of insecticides.

4.1. Application of Breeding for Pest Resistance

The genetics and inheritance of resistance to insect pests in crop species have been widely discussed by several researchers (Gatehouse et al., 1994; Panda and Khush, 1995; Smith, 2005). Wilson and George (1979) assessed the combined resistance of a group of cotton cultivars against pink bollworm (*Pectinophora gossypiella*). Estimates of the inheritability of resistance to *Trichoplusia ni* (Hubner) (Luedders and Dickerson, 1977) and *E. varivestis* (Sisson et al., 1976) pointed to quantitative inheritance in soybean.

Crossing between resistant parent and susceptible parent of *Pseudoplusia includens* in F_2 generation indicated the presence of partial dominance or quantitative inheritance (Kilen et al., 1977). The resistance of the lima bean (*Phaseolus lunatus* L.) to the leafhopper *Empoasca kraemeri* was found to be due to the quantitative effects of several genes and is a recessive trait (Lyman and Cardona, 1982). Additive and dominant genetic effects explain the resistance of the common bean (*Phaseolus vulgaris* L.) to *E. kraemeri (*Kornegay and Temple, 1986). Mehlenbacher et al., (1984) studied the heritability of lobed type A and type B trichome densities in *S. berthaultii* leaves, finding that resistance to aphid (*M. persicae*) was associated with the density of trichomes and the size of the droplets they exude; for this reason resistance was rated as a quantitative trait._Resistance to the potato tuber moth *Phtorimaea operculella* (Zeller) was found to be controlled by a small number of larger genes (Ortiz et al., 1990).

Recurrent selection was used to increase the levels of maize resistance to *Helicoverpa zea* and *Spodoptera frugiperda* (Butron et al., 2002). Maize spike and thatch resistance to the corn stalk borer *Sesamia nonagrioides* (Lefebvre) has also been found to be quantitatively inherited, and the interaction of dominant, additive, and epistatic effects controls the action of the genes (Cartea et al., 1999, Cartea et al., 2001). Additive and dominant effects explain a large part of the variation in maize resistance to the aphid *Rhopalosiphum maidis* (Bing and Guthrie, 1991) and the spotted stalk borer *Chilo partellus* (Swinhoe) (Pathak, 1991).

Advantages: Most important outcome of using the breeding approach is obtaining a cultivar with desired traits. The method is inexpensive to conduct as the base population can be a landrace. The population size selected can be variable, based on the objective. It is applicable to improving traits of low heritability, because selection is based on progeny performance.

Disadvantages: The purity of the cultivar may be hindered through admixture, natural crossing with other cultivars and mutations. Such off-type plants need to be omitted out to maintain cultivar purity. The cultivar has a narrow genetic base and, hence, is susceptible to devastation from adverse environmental factors because of uniform response. Instead of creating a new genotype, improvement is limited to isolation of the most desirable or best genotype from mixed population.

The method promotes loss in genetic pool because most superior pure lines are identified and multiplied for the exclusion of other genetic variants.

There are several disadvantages of these conventional methods such as the maintenance of cultivar purity, protection of environment, soil, and human health. To avoid the long time planning before sowing a crop, scientists had diverted their research interest from the conventional insect pest management strategies towards biotechnological strategies. Since 1980's these transgenic approaches came into practice and are still followed successfully.

5. Transgenic Approaches for Insect Resistance

The use of gene transfer technology to transfer insect resistance genes from diverse origins into crop plants provides an economical and eco-friendly alternative to the extensive use of toxic chemicals for the control of insect pests. Insect resistant transgenic plants may be produced by introducing foreign genes encoding either δ-endotoxin, protease inhibitors, lectins or amylase inhibitors (Gatehouse et al., 1997). The most widely used and well documented approach in this context is the insecticidal crystal protein (*Cry*) genes of *Bacillus thuringiensis* (*Bt*) coding insecticidal δ-endotoxin (Whitley and Schnepf, 1986). These toxins are highly specific to the target insects, non-toxic to animals and human beings, non-hazardous and eco-friendly (Schnepf, et al., 1998; Gatehouse, 2008).

Several strategies have been proposed for the management of resistance development in herbivorous field insects including the application of diverse mixture of toxins, high expression of *Bt* -toxin, weedy refugia, hybrid and pyramiding of different *Bt* -toxin genes and use of sterile insects (Gatehouse, 2008; Tabashnik et al., 2010). The *cry*1 proteins are classified based on amino acid sequence and the proteins designated *cry*1A (including *cry*1Aa, *cry*1Ab and *cry*1Ac) have greater than 85% identical amino acid sequence (Hofte and Whitely, 1989; Crickmore et al., 1998). Upon activation by proteases in the midgut, *Cry* toxin binds to more abundant but low-affinity glicosylphophatidylinositol

(GPI)-anchored protein alkaline phosphatase or aminopeptidase-N (ALP and APN, respectively) through domain II loop 3 and domain III ß16.

5.1. *Bt* Insecticidal Crystal Proteins

The gram-positive soil bacterium, *Bacillus thuringiensis* (*Bt*) produces proteinaceous crystalline inclusion bodies during sporulation which are effective against Lepidopteran, Dipteran, and Coleopteran group of agricultural insects. Crops engineered for crystal protein gene of *B. thuringiensis* have the added benefits of lower input and production costs.

B. thuringiensis produces an insecticidal crystal protein (ICP), which forms inclusion bodies of regular bipyramidal or cuboidal crystals during sporulation. The toxin genes are classified into four types, based on insect specificity and sequence homology. *Bt -cry* type I genes encode proteins of 130 kDa, and are usually specific to lepidopteran larvae, type II genes encode for 70kDa proteins that are specific to lepidopteran and dipteran larvae, type III genes encode for coleopteran larvae and the type IV genes are specific to the dipteran larvae.

The crystalline proteins of the bacterium get solubilized in alkaline mid gut of susceptible insects releasing activated proteins called δ-endotoxins.

The toxin portion is derived from the N-terminal half of the protoxin, while the C - terminal portion is involved in the formation of parasporal inclusion bodies and is hydrolyzed into small peptides. The region within N-terminal toxin domain (amino acids 1 to 27) is composed of α-helices, which are used for penetrating the peritrophic membrane. At least six α-helices can be identified on most of the Cry toxins.

The main target for *Bt* toxins is insect midgut. The crystalline protoxins are inactive until they are solubilized by the gut proteases, which cleave around 500 amino acids from the C-terminus of 130 kDa protoxin and 28 amino acids from the N-terminus leaving 55 protease resistant active core comprising the N-terminal half of the protoxin. Brush border membrane vesicles (BBMV) of susceptible insects have been identified as the primary binding site for insecticidal response. The active toxins bind to the specific receptors located on the apical brush border membrane of the columnar cells. There may be many toxin binding protein receptors, and some have been identified as 120 to180 kDa glycoproteins. After binding to the receptors, the toxin inserts irreversibly into plasma membrane of the cell, leading to lesion formation. After binding of toxin onto the midgut epithelial cells, the α-helices penetrate the apical membrane to form an ion-channel, allowing rapid flux of ions. The pores are K+ selective, permeable to cations, anions, or solutes such as sucrose, irrespective of the charge.

These pores possess both selective (only K+) and non-selective (Na+ and anions) properties depending on the pH, which eventually destroys the membrane potential.

Genes encoding for the δ-endotoxins have been cloned in 1980's, and expression of transgenes in tobacco and tomato produced the first genetically modified plants with resistance to insects. Subsequently a large number of agriculturally important crop plants expressing different *Bt* genes for resistance to specific agricultural insects have been developed expressing toxin upto 0.02% of the total soluble protein.

Truncation of gene, use of different promoters, enhancer sequences and fusion proteins resulted in only a marginal improvement in *Bt* -gene expression in plants. The expression level of native *Bt -cry* gene in plants was low to confer efficient protection against field insect pests. *B. thuringiensis cry* genes are typically A/T rich as compared to plant genes which are G/C rich.

Therefore prokaryotic *cry* gene sequences may contain transcription termination (polyadenylation) sites; *cry*ptic mRNA instability motifs (ATTA) that account for poor expression and stability of *Bt -cry* proteins in eukaryotic plant cells (Perlak et al., 1990, 1991; Murray et al., 1991). Modifying the *B. thuringiensis* sequences with synonymous codons to reduce or eliminate the potentially deleterious sequences resulted in improved expression to several folds (Fujimoto et al., 1993). Subsequently, several completely modified and synthetic *Bt -cry* genes have been developed for high-level expression in different plant species. Since the first successful transfer of *Bt -cry* genes in plant species, several crops including cotton, maize, rice, tomato, soybean, canola and sugarcane have been transformed with different *cry* gene and stable transgenic lines have been developed for field assessment.

Stable transgenic plants of highly recalcitrant crop plants like maize, cotton, and rice expressing one or several *cry* genes together have been well established. However, success of expressing *Bt -cry* genes in grain legumes and development of stable transgenic plants are only few.

Millions of small and large farmers both in industrialized and developing countries continue to increase their plantings of GM crops harbouring *Bt*-endotoxin gene because of the significant benefits they offer.

Bt is a highly sustainable and resource efficient crop management practice that requires less energy and conserves natural resources. It is more efficient in the control of insect pests and weeds. *Bt* kills only the harmful insects, and not the useful ones like other chemical pesticides. The second advantage of *Bt* is that it is non toxic to humans and wildlife. Therefore, it is used on many food crops and other sensitive crops where pesticides might harm the consumer, or local wildlife.

5.2. Plant Protease Inhibitors

Plant protease inhibitors (PIs), which play an important role against predators and pathogens, are natural defense-related proteins often present in seeds and induced in certain plant tissues by herbivore or wounding. The role of protease inhibitors (PIs) in plant protection was investigated in early 1947 when Mickel and Standish reported that the larvae of few insects were unable to develop normally on soybean. Afterwards, the trypsin inhibitors present in soybean were found out to be toxic to the larvae of flour beetle (*Tribolium confusum*) (Lipke et al., 1954). These protease inhibitor genes have benefits over genes coding for complex pathways e.g., by transferring single resistant gene from one plant species to another and expressing them from their constitutive promoters and imparting resistance against insect pests (Boulter, 1993).

This was first shown by Hilder et al. in 1987 by transferring trypsin inhibitor gene from *Vigna unguiculata* to tobacco, which bestowed resistance to wide range of insect pest orders including lepidopterans, coleopterans and orthoptera.

6. RNAi Approach to Develop Insect Free Plants

A method of gene silencing was developed in late 1990s on the basis of basic biological research in plants: RNA interference (RNAi). RNAi has been used extensively to genetically modify plants in the 2000s as plasmid vectors and more plant-specific biological information became available. The actual mechanism of RNAi was first discovered in nematodes (Fire et al., 1998). RNAi is now known to be a natural molecular pathway that all higher organisms use to defend themselves against parasites and pathogens.

For uses in GE crops, non coding RNA production in the form of double stranded RNA (dsRNA) is an efficient way to start a cascade of molecular events within the cell to silence genes in targeted plant gene or pest gene of interest, which can result in an insect resistance trait.

6.1. Application of RNAi in Insect Resistance

Most of the insects are voracious feeders during their larval stages. Thus researchers mostly target larval stage or before larval stage (pupae or neonate) for better resistance.

In case of *Helicoverpa armigera*, Cytochrome P450 gene of tobacco was targeted to cause retardation in larval growth (Mao et al., 2007). Similarly, HMGCoA reductase of cotton and chitinase of tomato were targeted for retardation in growth of larva in the same insect (Tian et al., 2015, Mamta et al., 2016).

Western corn rootworm (*Diabrotica virgifera*) and *Helicoverpa armigera* are also subjected to larval mortality through targeting v*ATPase*

A of maize and Ecdysone receptor gene of tobacco respectively (Baum et al., 2007, Zhu et al., 2012).

7. Genome Editing

CRISPR is the most recently developed tool for genome editing. Prokaryotes use CRISPR as an intrinsic line of defense against various viruses and plasmids by using RNA-guided nucleases to target and cleave foreign DNA sequences. The CRISPR/Cas system used in genome editing is primarily the Type II CRISPR/Cas9 isolated from *Streptococcus pyogenes*, in which foreign DNA sequences are introduced in between repeated sequences at the CRISPR locus and then transcribed into an RNA molecule known as crRNA (Sander and Joung, 2014). The crRNA then hybridizes with a second RNA (tracrRNA) and the complex binds to the Cas9 nuclease. The crRNA guides the complex to the target DNA which is then cleaved by the Cas9 nuclease. Researchers have dissected the innate CRISPR/Cas9 system and re-engineered it in such a way that a single RNA (guide RNA) is needed for Cas9-mediated cleavage of a target sequence. Guide RNA design requirements are limited to unique sequences of about 20 nucleotides in the genome and restricted near the protospacer adjacent motif sequence, which is specific for the CRISPR/Cas system. Newer applications of CRISPR include use of two unique guide RNAs with a modified nuclease that "nicks" one strand of the DNA, providing specificity for targeted deletions. The specificity of the guide RNA, and easiness of the CRISPR/Cas9 system have resulted in rapid utilization of this method of editing genomes in plants and other organisms (Baltes and Voytas, 2014). Thus, CRISPR/Cas9 system becomes an important future tool for insect resistance development cultivars.

8. WILD RELATIVES

Wild relatives of crops are ancestors of crop plants, having some inherent properties of defense against biotic and abiotic stresses in comparison to their cultivars, which habituate them to grow and develop in harsh environmental conditions. These inherent properties of wild relatives of crops plants are nowadays becoming the prime focus of scientists and researchers. Researchers are trying to explore the genes responsible for biotic and abiotic stress tolerant traits and transfer them to the cultivars to develop the variety having more resistance towards these stresses. It is an emerging technology in research and a very useful tool and has lesser regulatory concerns as the transgene is a plant-based gene from its own wild relative.

CONCLUSION

The primary advantage obtained from the ability to use genes from different organisms has been to increase the amount of genetic variability available or by providing a vast gene pool for breeders to use beyond that which were accessible by conventional breeding methods. The aim is to allow breeders to produce high yielding and productive crop varieties. Diversity in the gene pool can also be made by introducing useful genes from closely related species or via inducing mutations within the plant genome for better traits. Transgenesis has been a pertinent player in the development of insect resistant plants. The knowledge acquired from this kind of research can be applied in many research areas of plant science, not only for the creation of new crop varieties with novel traits, but also in the basic understanding of the various mechanisms. Still, scientists are working in the direction to develop foolproof methods for insect resistant crops with the help of plants' own defence mechanism and by making some alterations at the genomic level.

References

Baltes, N. J. and Voytas, D. F. (2014). Enabling plant synthetic biology through genome engineering. *Trends in Biotechnology,* 33:120 – 131.

Baum, J. A., Bogaert, T., Clinton, W., Heck, G. R., Feldmann, P., Ilagan, O., Johnson, S., Plaetinck, G., Munyikwa, T., Pleau, M., Vaughn, T. and Roberts, J. (2007). Control of coleopteran insect pests through RNA interference. *Nature Biotechnology,* 25:1322 – 1326.

Bing, J. W. and Guthrie, W. D. (1991) Generation mean analysis for resistance in maize to the corn leaf aphid Homoptera: Aphididae. *Journal of Economic Entomology*, 84:1080 – 1082.

Boulter, D. (1993). Insect pest control by copying nature using genetically engineered crops. *Biochemistry,* 34: 1453 - 1466.

Butron, A., Widstrom, N. W., Snook, M. E. and Wiseman, B. R. (2002) Recurrent selection for corn earworm Lepidoptera: Noctuidae resistance in three closely related corn southern synthetics. *Journal of Economic Entomology*, 95: 458 – 462.

Cartea, M. E., Malvar, R. A., Butrón, A., Vales, M. I. and Ordás, A. (1999) Inheritance of antibiosis to Sesamianonagrioides Lepidoptera: Noctuidae in maize. *Journal of Economic Entomology,* 92: 994 – 998.

Cartea, M. E., Malvar, R. A., Vales, M. I., Butrón, A. and Ordás, A. (2001) Inheritance of resistance to ear damage caused by Sesamianonagrioides Lepidoptera: Noctuidae in maize. *Journal of Economic Entomology,* 94: 277 – 283.

Crickmore, N., Zeigler, D. R., Feitelson, J., Schnepf, E., Van Rie, J., Lereclus, D., Baum, J. and Dean, D. H. (1998). Revision of the nomenclature for the *Bacillus thuringiensis* pesticidal crystal proteins. *Microbiology and Molecular Biology Reviews*, 62: 807 – 813.

Fire, A. Xu, S., Montgomery, M. K., Kostas, S. A., Driver, S. E., and Mello, C. C. (1998). Potent and specificc genetic interference by double-stranded RNA in *Caenorhabditis elegans*. *Nature,* 391: 806 – 811.

Fordyce, J. A. 2001. The lethal plant defense paradox remains: inducible host plant aristolochic acids and the growth and defense of the pipevine swallowtail. *Entomologia Experimentalis et Applicata,* 100: 339 – 3.

Fujimoto, H., Itoh, K., Yamamoto, M., Kyozuka, J. and Shimamoto, K. 1993. Insect resistant rice generated by introduction of a modified delta-endotoxin gene of *Bacillus thuringiensis*. *Bio. Technology*, 11: 1151 - 1155.

Gallo, D., Nakano, O., Neto, S. S., Carvalho, R. P. L., Batista, G. C., Filho, E. B., Parra, J. R. P., Lopes, J. R. S. and Omoto, C. (2002). Entomologia Agrícola. FEALQ, Piracicaba.

Gatehouse, A. M., Boulter, D. and Hilder, V. A. (1994). Potential of plant-derived genes in the genetic manipulation of crops for insect resistance. *Plant Genetic Manipulation for Crop Protection*, 7: 155 – 181.

Gatehouse, J. A. (2002). Plant resistance towards insect herbivores: a dynamic interaction. *New Phytologist*, 156: 145 – 169.

Gatehouse, J. A. (2008). Biotechnological prospects for engineering insect-resistant plant. *Plant Physiology*, 146: 881 – 887.

Gatehouse, L. N., Shannon, A. L., Burgess, E. P. J. and Cristeller, J. T. (1997). Characterization of major midgut protease cDNAs from *Helicoverpa armigera* larvae and expression in response to four proteinase inhibitors in diet. *Insect Biochemistry and Molecular Biology*, 27: 929 - 944.

Hilder, V. A., Barker, R. F., Samour, R. A., Gatehouse, A. M. R., Gatehouse, J. A. and Boulter, D. (1989). Protein and cDNA sequences of Bowman-Birk protease inhibitors from the cowpea (*Vigna unguiculata* Walp). *Plant Molecular Biology,* 13(6): 701 - 710.

Hofte, H. and H. R. Whiteley. (1989). Insecticidal crystal proteins of *Bacillus thuringiensis*. *Microbiological Reviews*, 53: 242 – 255.

Kilen, T. C., Hatchett, J. H. and Hartwig, E. E. (1977). Evaluation of early generation soybeans for resistance to soybean looper. *Crop Science*, 17: 397 – 398.

Kornegay, J. L. and Temple, S. R. (1986). Inheritance and combining ability of leafroller defense mechanisms in common bean. *Crop Science*, 26:1153 – 1158.

Lipke, H., Fraenkel, G. S. andLiener, I. E. (1954). Growth inhibitors. Effect of soybean inhibitors on growth of *Tribolium confusum*. *Journal of Agricultural and Food Chemistry*, 2: 410 - 414.

Luedders, V. D. and Dickerson, W. A. (1977). Resistance of selected soybean genotypes and segregating populations to cabbage looper feeding. *Crop Science*, 17: 395 – 396.

Lyman, J. M. and Cardona, C. (1982). Resistance in lima beans to a leafhopper, *Empoasca krameri*. *Journal of Economical Entomology*, 75: 281 – 286.

Mamta, Reddy, K. R. and Rajam, M. V. (2016). Targeting chitinase gene of *Helicoverpa armigera* by host-induced RNA interference confers insect resistance in tobacco and tomato. *Plant Molecular Biology,* 90: 281 – 292.

Mao, Y. B., Cai, W. J., Wang, J. W., Hong, G. J., Tao, X. Y., Wang, L. J., Huang, Y. P. and Chen, X. Y. (2007). Silencing a cotton bollworm P450 monooxygenase gene by plant-mediated RNAi impairs larval tolerance of gossypol. *Nature Biotechnology*, 25: 1307 – 1313.

Mehlenbacher, S. A., Plaisted, R. L. and Tingey, W. M. (1984). Heritability of trichome density and droplet size in interspecific potato hybrids and relationship to aphid resistance. *Crop Science*, 24: 320 – 322.

Murray, E. E., Rocheleau, T., Eberle, M., Stock, C., Sekar, V., and Adang, M. (1991). Analysis of unstable RNA transcripts of insecticidal crystal protein genes of Bacillus thuringiensis in transgenic plants and electroporated protoplasts. *Plant Molecular Biology*, 16(6): 1035 - 1050.

Ortiz, R., Iwanaga, M., Raman, K. V. and Palacios, M. (1990). Breeding for resistance to potato tuber moth, *Phthorimaea operculella* Zeller in diploid potatoes. *Euphytica,* 50: 119 – 125.

Panda, N. and Khush, G. S. (1995) Host plant resistance to insects. Guildford, Biddles.

Pathak, R. S. (1991). Genetics of sorhgum, maize, rice and sugarcane resistance to the cereal stem borer Chilo spp. *Insect Science And Its Application*, 11: 689 – 699.

Perlak, F. J., Fuchs, R. L., Dean, D. A., McPherson, S. L. and Fischhoff, D. A. (1991). Modification of the coding sequence enhances plant expression of insect control protein genes. *Proceedings of the National Academy of Sciences USA*, 88: 3324 – 3328.

Perlak, F. J., Deaton, R. W., Armstrong, T. A., Fuchs, R. L., Sims, S. R., Greenplate, J. T. and Fischhoff, D. A. (1990). Insect resistant cotton plants. *Biotechnology*, 8: 939 – 943. 295.

Sander, J. D. and Joung. J. K. (2014). CRISPR-Cas systems for editing, regulating and targeting genomes. *Nature Biotechnology*, 32: 347 – 355.

Schnepf, E., Crickmore, N., Van Rie, J., Lereclus, D., Baum, J., Feitelson, J., Zeigler, D. R. and Dean, D. H. (1998). *Bacillus thuringiensis* and its pesticidal crystal proteins. *Microbiology and Molecular Biology Reviews,* 62: 775 – 806.

Sisson, V. A., Miller, P. A., Campbell, W. V. and Van Duyn, J. W. (1976). Evidence of inheritance of resistance to the Mexican bean beetle in soybeans. *Crop Science*, 16: 835 – 837.

Smith, C. M. (2005). Plant resistance to arthropods–molecular and conventional approaches. Springer, New York.

Smith, R. F. (1970). Pesticides: their use and limitations in pest management. *In*: Rabb, R. L. and Guthrie, F. E. (eds) Concepts of pest management. North Caroline State University, Raleigh. pp. 103 - 118.

Tabashnik, B. E., Sisterson, M. S., Ellsworth, P. C., Dennehy, T. J., Antilla, L., Liesner, L., Whitlow, M., Staten, R. T., Fabrick, J. A., Unnithan, G. C., Yelich, A. J., Ellers-Kirk, C., Harpold, V. S., Li, X. and Carriere, Y. (2010). Suppressing resistance to Bt cotton with sterile insect release. *Nature Biotechnology,* 28: 1304 – 1307.

Whiteley, H. R. and Schnepf, H. E.. (1986). The molecular biology of parasporal crystal body formation in *Bacillus tlhurintgienisis*. *Annual Review of Microbiology*, 40: 549 - 576.

Wilson, F. D. and George, B. W., (1979), Combining ability in cotton for resistance to pink bollworm. *Crop Science*, 19: 834 – 836.

Zhu, J. Q., Liu, S., Ma, Y., et al., Improvement of pest resistance in transgenic tobacco plants expressing dsRNA of an insect-associated gene *EcR*. Doucet D, ed. *PLoS ONE*, 2012; 7(6):e38572.

In: Functional Genomics ISBN: 978-1-53612-565-8
Editors: Holger Uffe and Olaf Philip

Chapter 4

THE EXPLOITATION OF INSECT RESISTANT TRAITS IN WILD RELATIVES AND THEIR UTILITY IN CROP IMPROVEMENT

Pragya Mishra[1], Maniraj Rathinam[1], M. K. Ramkumar[1], Shweta Singh[1], Ravi Prakash Saini[1], Kesiraju Karthik[1], Suraj Singh[1], N. Muralimohan[1], T. Arulprakash[1], Nikhil Ramkumar[1], Saurabh Tyagi[1], Era Vaidya Malhotra[1], T. Vinutha[2], Debasis Pattanayak[1] and Rohini Sreevathsa[1,*]

[1]ICAR- National Research Center on Plant Biotechnology, LBS building, Pusa Campus, New Delhi, India
[2]Division of Biochemistry, ICAR-Indian Agricultural Research Institute, New Delhi, India

[*] Corresponding author Email: rohinisreevathsa@gmail.com.

Abstract

Wild relatives of crops (WRC) are the ancestors of crop plants and other species closely related to crops, which are seen to be continuously growing under harsh conditions. Wild relatives of plants are differentially exposed to nature and live by different rules of adaptability and survival than their cultivated counterparts. Continuous exposure to severe climatic conditions has helped them evolve at the genetic level, making them more genetically diverse. These wild species are considered as one of the most important sources of resistance genes to tackle both biotic and abiotic stresses. Due to their broader genetic base, wild species have a potential role in the development of crops which can withstand adverse impacts of climate change, such as increasing scarcity of nutrients, water and other inputs, and increased incidents of new pests and diseases. These vital genetic resources can be secured for future use. Various morphological, physiochemical, biochemical and molecular mechanisms are responsible for making wild crop relatives resistant to various stresses. In this chapter the important biotic stress responsive traits possessed by wild relatives, specifically against insect pests, and their potential uses in crop improvement programmes are discussed.

Keywords: insect resistance, wild relatives of crops, host defense response, biotic stress, genetic variation

1. Introduction

Insect resistance is a major concern in crop improvement. Insect herbivore attack causes great devastation of crop plants thereby leading to adverse effects on crop yield (Oerke, 2006). To overcome this problem, techniques of genetic engineering were adopted for crop improvement (Ramu et al., 2012). But due to high selection pressure insects have evolved and developed resistance (Kumari and Kumar, 2015). This scenario has led to the continued search for novel traits to mitigate pest attack on plants. Wild Relatives of Crops (WRC) are reported to have a comparatively broader spectrum of resistance than the cultivated varieties. Increased larval mortality rate, stunted larval growth and hindrance to oviposition have been observed when insects infest the WRCs due to morphological and biochemical barriers present in them. Exploiting WRC and non-host crop plants for insect resistance is a very effective way to overcome the problem of insect herbivores (Harsulkar et al., 1999; Sharma et al., 2009). Pest resistance traits vary between plant species, and this variation could be due to the genetic makeup (Bull and Wichman, 2001), natural selection pressure (Mooney et al., 2010) and domestication effects. Domestication (human artificial selection) causes evolution of crop plants having selected agronomical traits. These trait-based selection procedures reduce genetic diversity in crop populations in comparison to the wild types (Chaudhary, 2013). Hence, in the recent times the focus of scientists globally has been to decipher the molecular mechanisms underlying resistance to insect pests in the wild relatives. Omics tools, such as, RNA seq, microarray technology, global transcriptomics and metabolome profiling are now being used to compare the cultivated and the wild relatives of a particular crop species for elucidating the underlying resistance/susceptibility mechanisms (Ehlting et al., 2008; Huang et al., 2015). Comparative global gene expression profiles of wild and cultivated cotton have shown downregulation of many defense responsive genes in the cultivated cotton species (Chaudhary et al., 2008). This loss of function can be due to mutation in regulatory elements during the domestication process or loss of loci during trait based selection (Hyten et al., 2006). Natural selection pressure accumulates resistance traits in WRC and these traits may be utilized towards crop improvement by both breeding and transgenic approaches. Developing hybrids with enhanced defense mechanisms through conventional breeding can be achieved only between closely related and compatible species (Mallikarjuna et al., 2007). On the other hand, in transgenic approach genes conferring resistance from distant species can also be utilized. In this chapter insect resistance in WRC has been discussed along with the strategies that could be used for exploitation of these traits for developing crops resistant to insect herbivore attack.

2. Morphological Barriers That Aid in Conferring Resistance to Insects

The foremost defense mechanism of plants against insect herbivores is the presence of structural barriers, broadly classified as Spinescence (spines, thorns), Sclerophyly (hardened leaves), and Pubescence (trichomes) (War et al., 2012a). Spinescence includes structures such as spines, thorns, prickles etc. that are reported to protect plants against a wide range of pests (Hanley et al., 2007). Hardening of leaves (sclerophyly) by cuticular wax accumulation, cell wall thickening and lignification prevents plants from insect attack by making them non-preferable, unpalatable and undigestable (White and Eigenbrode, 2000; Hanley et al., 2007; Sujana et al., 2008). Pubescence comprises of the hairy layer protruding out from the epidermal layer known as trichomes; which may be present in different forms *viz;* spiral, branched, unbranched, hooked, glandular and non-glandular (Hanley et al., 2007; Sharma et al., 2009). Glandular trichomes play a critical role in defense mechanism, as they are not just physical barriers but are also reported to secrete chemical compounds like alkaloids, terpenoids, flavonoids which are poisonous or repellent or in some cases act as traps against insects and other organisms (Hanley et al., 2007). This phenomenon is positively correlated to insect resistance in wild relatives of pigeon pea, where the glandular and the non glandular trichomes present on the pod surface have been reported to be defensive against *Helicoverpa armigera* (Sharma et al., 2009).

3. Induced Defense Responses to Insect Attack

Plant defense system is a highly complex and biologically conserved process. It depends on two important factors, growth stage of plant and pest abundance. When the plant is in early vegetative stage, it selects defensive approach, whereas at reproductive stage the plant opts for stress avoidance (Rhoades and Cates, 1976; Bazzaz et al., 1987; Stamp, 2003).

When insect herbivores start feeding, plants recognize the fatty acid - amino acid conjugates (FACs) from insect's salivary secretion and triggers plant elicitors which transduce information to neighboring plants and other predator species (Turlings et al., 1993). Simultaneously, it also activates MAPK (Mitogen Activated Protein Kinase) which is the central hub for various biotic and abiotic stress responses (Ryan, 2000). MAPK activates three major phytohormones: jasmonic acid, salicylic acid and ethylene (Furstenberg et al., 2013). These hormones are involved in the activation of various defense regulatory genes that are grouped as toxic, antinutrient and unpalatable (Furstenberg et al., 2013).

The induced defense responses are mostly controlled by single genes which follow different modes of action (Table1). A particular group of enzymes act on the nutritional components and makes them unavailable. Enzymes like argenase, theronine deaminase and asparginase degrade the specific essential amino acids present in the insect mid gut (Chen et al., 2004; Chen et al., 2005). The peroxidase and polyphenol oxidase convert phenolic compounds to quinines and other oxidative radicals. These derivatives bind with the plant proteins and reduce their nutrition quality (Stout et al., 2009; Bhonwong et al., 2009). Lipoxygenase is involved in both signaling and direct response. Hydroperoxidation of polyunsaturated fatty acids by lipoxygenase leads to formation of fatty acid hydroperoxidase, which undergoes enzymatic cleavage resulting in the formation of unstable products like aldehydes, γ-ketoles, epoxides and other reactive oxygen species (Bruinsma et al., 2009). These lead to the depletion of amino acids by forming protein-protein cross linking (Bruinsma et al., 2009).

Protease inhibitors are reported to be expressed in remarkably high levels in response to pest attack across a wide range of plant species (Parde et al., 2012). Diverse classes of digestive enzymes are inhibited by these protease inhibitors, which limit the nutrient utilization leading to retarded larval growth (Parde et al., 2012).

Table 1. Induced plant defense protease inhibitors and enzymes against insect herbivore attack

Protein and enzymes	Plant sources	Insect species	Reference
Protease Inhbitors	*Sorghum bicolor*	*Manduca sexta*	Zhu et al., 2004
	Tomato	*Helicoverpa armigera*	Chen et al., 2005
	Solanum nigrum	*Spodoptera littoralis*	Hartl et al., 2010
	Nicotiana attenuata	*Spodoptera exigua*	Steppuhn et al., 2007
	Arabidopsis	*Plutella xylostella*	De et al., 2001
Superoxide dismutase	*Medicago sativa*	*Aphis medicaginis*	Huang et al., 2007
Oxidative enzymes			
Polyphenol oxidases	Tomato	*Manduca sexta*	Chen et al., 2005
	Buffalograss	*Blissus oxiduus*	Heng et al., 2004
	Tomato	*Spodoptera frugiperda, Helicoverpa armigera*	Bhonwong et al., 2009
Catalase	Bufallograssses	*Blissus oxiduus*	Heng et al., 2004
Peroxidases	*Alnus glutinosa*	*Agelastica alni*	Tscharntke et al., 2001
	Arabidopsis	*Bemisia tabaci*	Kempema et al., 2007
	Poplar	*Lymantria dispar*	Huang et al., 2007
	Medicago sativa	*Aphis medicaginis*	Chen et al., 2009
	Corn	*Spodoptera littoralis*	Stout et al., 2009
Lipoxygenases	*Cucumis sativus*	*Spodoptera littoralis*	Reymond et al., 2004
	Nicotiana attenuata	*Bemisia tabaci*	Kempema et al., 2007
	Alnus glutinosa	*Agelastica alni*	Tscharntke et al., 2001
	Wheat	*Sitobion avenae*	Zhao et al., 2009
	Tomato	*Macrosiphium euphorbiae, Myzus persicae*	Fidantsef et al., 1999
Amino acid deaminases	Tomato	*Manduca sexta*	Chen et al., 2005
	Tomato	*Manduca sexta*	Kang et al., 2006
	Wheat	*Mayetiola destructor*	Liu et al., 2007

Intestinal cell wall disruption is another type of defense response (Table 2). Lectins are glycoproteins that are highly stable in a broad range of pH conditions and are also resistant to digestive enzymes (Vandenborre et al., 2010). This feature makes lectin a potential insecticide. Lectin binds to the insect intestinal tract and interferes with nutrient absorption leading to a systemic reaction in insects (Chakraborti et al., 2009). Chitinases are enzymes that target chitins present on the peritrophic membrane in insects. Overexpression of chitinase in tomato has been reported to increase resistance against *Leptinotarsa decemlineata* (Lawrence and Novak, 2006). Cysteine protease is a papain like protease that gets agglomerated at the feeding site. Post ingestion, these proteases digest the proteins present in the gut region damaging the peritrophic membrane thereby inhibiting insect growth (Pechan et al., 2000). These strategies of defense and their association with high level of resistance are reported in various WRC (Sharma et al., 2009). In wild relatives of pigeon pea, accessions expressing a wide range of protease inhibitors are found to be polymorphic in their size, number and functional specificity towards their corresponding proteases and have shown significant inhibition against different proteases of *Helicoverpa armigera* than the cultivars (Parde et al., 2012).

Table 2. Induced plant defense proteins and enzymes against insect herbivore attack

Protein and enzymes	Plant sources	Insect species	Reference
Lectin like proteins	Wheat	*Nilaparvata lugens*	Saha et al., 2006
	Chickpea	*Aphis craccivora*	Chakraborti et al., 2009
	Tobacco	*Spodoptera littoralis, Manduca sexta,*	Vandenborre et al., 2009
	Arabidopsis	*Aphis craccivora*	Majumder et al., 2005
	Tobacco	*Anagasta kuehniella*	Macedo et al., 2007
Chitinases	*Sorghum bicolor*	*Schizaphis graminum*	Zhu et al., 2004
	Phaseolus vulgaris	*Lacanobia oleracea*	Gatehouse et al., 1997
	Populus spp	*Leptinotarsa decemlineata*	Lawrence and Novak, 2006
Protease	Corn	*lepidopterans*	Pechan et al., 2002
	Papaya	*Samia ricini, Mamestra brassicae, Spodoptera litura*	Konno et al., 2004
	Tomato	*Spodoptera littoralis*	Pautot et al., 1993

4. Natural Variation as a Plant Defense Mechanism

The intensity of plant defense response varies across species. In case of pigeon pea wild relatives, different accessions from the same species have demonstrated a range of responses against *Helicoverpa armigera* (Sharma et al., 2009; Parde et al., 2012). This variation could be mainly due to gene diversity, sequence polymorphism or expression level polymorphisms. The survival and adaptability of plant species to different biotic and abiotic stresses is due to genetic diversity (Mooney et al., 2010). This variation is acquired from their origin and other influencing factors to which these species are subjected (Mooney et al., 2010). Another important contributing factor could be the variation in structural genes and regulatory elements leading to sequence level and expression level polymorphisms. These variations in genotypes result in changes leading to effective defense response (Kliebenstein, 2009).

Level of gene expression could directly affect the level of defense response against insect herbivores. Next generation sequencing technologies and Quantitative PCR analysis are helpful in finding the expression level variation (Ehlting et al., 2008). Several studies have demonstrated this globally. Induced expression of PPO in groundnut and tomato has shown significant resistance against *Helicoverpa armigera*, supporting the aforementioned fact (Bhonwong et al., 2009; War et al., 2012b).

Epigenetic variations in gene sequences also affect gene expression. Methylation-sensitive amplified polymorphism (MSAP) is an effective tool to identify variation in methylation pattern (Pandey et al., 2017). In foxtail millet, salt stress has been studied to cause variation in the methylation sites (Pandey et al., 2017). Genome wide methylation variation in response to herbivore attack is reported in *Taraxacum officinale* (Verhoeven et al., 2010). Identifying the natural variation linked to high level resistance helps in finding efficient genes for crop improvement against insect herbivore attack.

5. Genetic Engineering for the Development of Insect Resistant Crops

5.1. Transgenic Strategies for Insect Resistance

Natural variations in insect resistant phenotypes were generally classified as qualitative and quantitative traits. Quantitative traits are often controlled by the multiple genes, whereas qualitative traits are commonly single gene traits (Yencho et al., 2000; Keurentjes et al., 2008). Insect resistance traits fall under the qualitative traits. Various insect resistance genes validated by transgenic approach are listed in Table 3. Genes or gene products involved in insect resistance have different modes of action. Over expressing multiple genes conferring resistance rather than using a single gene provides enhanced resistance due to synergetic effect (Duffey et al., 1996: Steppuhn et al., 2007). In the case of tomato, protease inhibitors, oxidative enzymes and secondary metabolites show significant resistance due to cumulative effect, but in the individual form they are less effective (Duffey et al., 1996). Expression of protease inhibitor and nicotine together in *Nicotiana attenuate* produced a synergetic effect on *Spodoptera exigua* (Steppuhn et al., 2007). Exploiting resistance traits from different host plants also helps in crop improvement (Table 3). Protease inhibitors from the non- host plants such as *Arachis hypogea*, *Psophocarpus tetragonolobus*, and *Solanum tuberosum* show effective inhibition of digestive proteases secreted in *Helicoverpa armigera* as compared to host plants *viz Cicer arietinum, Cajanus cajan* and *Gossypium arboretumv* (Harsulkar et al., 1999). Lipoxygenase is directly involved in defense mechanism along with its role in the formation of certain chemical compounds that provides information for the predator species (Mao et al., 2007; Bruinsma et al., 2009). A combination of direct defense genes from other species along with lipoxygenase could help in long term protection. Using the transgenic approach, it is possible to employ different traits from different plant species together into one plant. One can utilize genes from any gene pool to develop a combination of resistance traits. Moreover this approach is less time consuming as compared to the conventional techniques.

Table 3. Transgenic crops developed by using host and non – host insect resistance traits

Defense gene products	Transgenic plants	Name of herbivore	Reference
Protease inhibitors			
Cowpea trypsin inhibitor	Tobacco	*Manduca sexta*	Hilder et al.,1987
	Cotton	*Helicoverpa armigera*	Charity et al.,1999
	Wheat	*Sitotroga cerealella*	Bi et al., 2006
Potato trypsin inhibitor	Tobacco	*Chrysodeixis reen*	McManus et al.,1994
Sweet potato trypsin inhibitor	Cauliflower	*Plutella xylostella*	Ding et al., 1998
Barley trypsin inhibitor Cme	Rice	*Sitophilus oryzae*	Alfonso et al., 2003
Common bean α - amylase inhibitor 1	Azuki bean	*Callosobruchus maculates*	Ishimoto et al., 1996
	Pea	*Bruchus pisorum*	Morton et al., 2000
	Chickpea	*Callosobruchus maculates*	Sarmah et al., 2004
Protease inhibitors			
Soybean Kunitz proteinase inhibitor	Poplar	*Lymantria dispar*	Confalonieri et al., 1998
Rice cystein protease inhibitor	Poplar	*Chrysomela tremulae*	Leplé et al., 1995
Tobacco mulit-domain inhibitor	Apple	*Epiphyas postvittana*	Maheswaran et al., 2007
Lectin like proteins			
Galanthus nivalis agglutinin	Rice	*Nilaparvata lugens*	Rao et al.,1998
	Rice	*Sogatella furcifera*	Ramesh et al., 2004
Wheat germ agglutinin	Indian mustard	*Lipaphis erysimi*	Kanrar et al., 2002
Allium sativam agglutinin	Rice	*Nilaparvata lugens*	Chandrasekhar et al., 2014
Polyphenol oxidase			
Tomato poly phenol oxidase	Tomato	*Helicoverpa armigera, Spodoptera exigua*	Bhonwong et al., 2009
Chitinases			
Common Bean chitinase	Potato	*Lacanobia oleracea*	Gatehouse et al., 1997
poplar chitinase	Tomato	*Leptinotarsa decemlineata*	Lawrence and Novak, 2006

5.2. Genome Editing for Insect Resistance

The Clustered Regularly Interspaced Short Palindromic Repeats (CRISPR)/Cas9 is an adaptive immune system present in the bacteria (Khatodia et al., 2016). It contains two major components *viz.,* CRISPR loci present in the genome and a Cas9 nuclease. CRISPR/cas complex comprise of a non coding RNA and a trans-activated RNA which are coupled to form a guide RNA. This guide RNA recognizes the target/foreign DNA sequence and Cas9 enodonuclease cleaves the double stranded target DNA (Kim and Kim, 2014). Understanding the CRISPR/Cas mechanism has led the way for the development of engineered RNA guided endonuclease system (Kim and Kim, 2014). This engineered CRISPR/Cas9 system can be utilized for site directed mutagenesis accordingly, based on the type of modification required (Andolfo et al., 2016).

A high degree of variation in insect resistant traits has been observed among different plant species (Harsulkar et al., 1999; Sharma et al 2009). Identifying these natural variations and altering the target gene sequence will provide a sustainable resistant capacity to the crop. CRISPR/Cas system makes precise editing in the genome and also at multiple target sites (Cong et al., 2013). Developing marker-free genetically edited crops through CRISPR/Cas can be used as a solution for social acceptance and regulatory issues.

Conclusion

The insect resistance efficiency in most of the cultivars currently used is comparatively low. Screening of more than 14,000 accessions of cultivated pigeonpea revealed moderate or low levels of resistance towards *Helicoverpa armigera.* To overcome this issue and to manage the adverse effects of insect herbivore attack without disturbing the environmental balance, the WRC population have to be exploited for their defense responsive mechanisms. Development of insect resistant crops has so far been done utilizing foreign genes rather than genes from WRC. The modern comparative omics approaches will help to take a closer look at the elite genes conferring resistance in WRC. Those wild relatives with promising resistance mechanism can be used as donors for transferring resistance traits to cultivars.

References

Alfonso, R. J., Ortego, F., Castañera, P., Carbonero, P. & Díaz, I. (2003). Transgenic expression of trypsin inhibitor CMe from barley in indica and japonica rice, confers resistance to the rice weevil *Sitophilus oryzae*. *Transgenic Research*, 12(1): 23-31.

Andolfo, G., Iovieno, P., Frusciante, L. & Ercolano, M. R. (2016). Genome-editing technologies for enhancing plant disease resistance. *Frontiers in Plant Science*, 7: 1813.

Bazzaz, F. A., Chiariello, N. R., Coley, P. D. & Pitelka, L. F. (1987). Allocating resources to reproduction and defense. *BioScience*, 37(1): 58-67.

Bhonwong, A., Stout, M. J., Attajarusit, J. & Tantasawat, P. (2009). Defensive role of tomato polyphenol oxidases against cotton bollworm (*Helicoverpa armigera*) and beet armyworm (*Spodoptera exigua*). *Journal of Chemical Ecology*, 35(1): 28-38.

Bi, R. M., Jia, H. Y., Feng, D. S. & Wang, H. G. (2006). Production and analysis of transgenic wheat (*Triticum aestivum* L.) with improved insect resistance by the introduction of cowpea trypsin inhibitor gene. *Euphytica*, 151(3): 351-360.

Bruinsma, M., Posthumus, M. A., Mumm, R., Mueller, M. J., van Loon, J. J. & Dicke, M. (2009). Jasmonic acid-induced volatiles of *Brassica oleracea* attract parasitoids: effects of time and dose, and comparison with induction by herbivores. *Journal of Experimental Botany*, 60(9): 2575-2587.

Bull, J. J. & Wichman, H. A. (2001). Applied evolution. *Annual Review of Ecology and Systematics*, 32(1): 183-217.

Chakraborti, D., Sarkar, A., Mondal, H. A. & Das, S. (2009). Tissue specific expression of potent insecticidal, Allium sativum leaf agglutinin (ASAL) in important pulse crop, chickpea (*Cicer arietinum* L.) to resist the phloem feeding *Aphis craccivora. Transgenic Research*, 18(4): 529-544.

Chandrasekhar, K., Vijayalakshmi, M., Vani, K., Kaul, T. & Reddy, M. K. (2014). Phloem-specific expression of the lectin gene from *Allium sativum* confers resistance to the sap-sucker *Nilaparvata lugens. Biotechnology Letters*, 36(5): 1059-1067.

Charity, J. A., Anderson, M. A., Bittisnich, D. J., Whitecross, M. & Higgins, T. J. V. (1999). Transgenic tobacco and peas expressing a proteinase inhibitor from *Nicotiana alata* have increased insect resistance. *Molecular Breeding*, 5(4): 357-365.

Chaudhary, B. (2013). Plant domestication and resistance to herbivory. *International Journal of Plant Genomics*, 1-14.

Chaudhary, B., Hovav, R., Rapp, R., Verma, N., Udall, J. A. & Wendel, J. F. (2008). Global analysis of gene expression in cotton fibers from wild and domesticated *Gossypium barbadense. Evolution & Development*, 10(5): 567-582.

Chen, H., McCaig, B. C., Melotto, M., He, S. Y. & Howe, G. A. (2004). Regulation of plant arginase by wounding, jasmonate, and the phytotoxin coronatine. *Journal of Biological Chemistry*, 279(44): 45998-46007.

Chen, H., Wilkerson, C. G., Kuchar, J. A., Phinney, B. S. & Howe, G. A. (2005). Jasmonate-inducible plant enzymes degrade essential amino acids in the herbivore midgut. *Proceedings of the National Academy of Sciences of the United States of America*, 102(52): 19237-19242.

Chen, Y., Ni, X. & Buntin, G. D. (2009). Physiological, nutritional, and biochemical bases of corn resistance to foliage-feeding fall armyworm. *Journal of Chemical Ecology*, 35(3): 297-306.

Confalonieri, M., Allegro, G., Balestrazzi, A., Fogher, C. & Delledonne, M. (1998). Regeneration of *Populus nigra* transgenic plants expressing a Kunitz proteinase inhibitor (KTi3) gene. *Molecular Breeding*, 4(2): 137-145.

Cong, L., Ran, F. A., Cox, D., Lin, S., Barretto, R., Habib, N., Hsu, P. D., Wu, X., Jiang, W., Marraffini, L. A. & Zhang, F. (2013). Multiplex genome engineering using CRISPR/Cas systems. *Science*, 339(6121): 819-823.

De, L. F., Bonadé, B. M., Ceci, L. R., Gallerani, R. & Jouanin, L. (2001). Effects of a mustard trypsin inhibitor expressed in different plants on three lepidopteran pests. *Insect Biochemistry and Molecular Biology*, 31(6): 593-602.

Ding, L. C., Hu, C. Y., Yeh, K. W. & Wang, P. J. (1998). Development of insect-resistant transgenic cauliflower plants expressing the trypsin inhibitor gene isolated from local sweet potato. *Plant Cell Reports*, 17(11): 854-860.

Duffey, S. S. & Stout, M. J. (1996). Antinutritive and toxic components of plant defense against insects. *Archives of Insect Biochemistry and Physiology*, 32(1): 3-37.

Ehlting, J., Chowrira, S. G., Mattheus, N., Aeschliman, D. S., Arimura, G. I. & Bohlmann, J. (2008). Comparative transcriptome analysis of *Arabidopsis thaliana* infested by diamond back moth (*Plutella xylostella*) larvae reveals signatures of stress response, secondary metabolism, and signalling. *BMC Genomics*, 9(1): 154.

Fidantsef, A. L., Stout, M. J., Thaler, J. S., Duffey, S. S. & Bostock, R. M. (1999). Signal interactions in pathogen and insect attack: expression of lipoxygenase, proteinase inhibitor II, and pathogenesis-related protein P4 in the tomato, *Lycopersicon esculentum*. *Physiological and Molecular Plant Pathology*, 54(3-4): 97-114.

Fürstenberg, H. J., Zagrobelny, M. & Bak, S. (2013). Plant defense against insect herbivores. *International Journal of Molecular Sciences*, 14(5): 10242-10297.

Gatehouse, A. M., Davison, G. M., Newell, C. A., Merryweather, A., Hamilton, W. D., Burgess, E. P. & Gatehouse, J. A. (1997). Transgenic potato plants with enhanced resistance to the tomato moth, *Lacanobia oleracea*: growth room trials. *Molecular Breeding*, 3(1): 49-63.

Hanley, M. E., Lamont, B. B., Fairbanks, M. M. & Rafferty, C. M. (2007). Plant structural traits and their role in anti-herbivore defence.

Perspectives in Plant Ecology, Evolution and Systematics, 8(4): 157-178.

Harsulkar, A. M., Giri, A. P., Patankar, A. G., Gupta, V. S., Sainani, M. N., Ranjekar, P. K. & Deshpande, V. V. (1999). Successive use of non-host plant proteinase inhibitors required for effective inhibition of *Helicoverpa armigera* gut proteinases and larval growth. *Plant Physiology*, 121(2): 497-506.

Hartl, M., Giri, A. P., Kaur, H. & Baldwin, I. T. (2010). Serine protease inhibitors specifically defend *Solanum nigrum* against generalist herbivores but do not influence plant growth and development. *The Plant Cell*, 22(12): 4158-4175.

Heng-Moss, T., Sarath, G., Baxendale, F., Novak, D., Bose, S., Ni, X. & Quisenberry, S. (2004). Characterization of oxidative enzyme changes in buffalo grasses challenged by *Blissus occiduus*. *Journal of Economic Entomology*, 97(3): 1086-1095.

Hilder, V. A., Gatehouse, A. M., Sheerman, S. E., Barker, R. F. & Boulter, D. (1987). A novel mechanism of insect resistance engineered into tobacco. *Nature*, 330(6144): 160-163.

Huang, X. Z., Chen, J. Y., Xiao, H. J., Xiao, Y. T., Wu, J., Wu, J. X. & Guo, Y. Y. (2015). Dynamic transcriptome analysis and volatile profiling of *Gossypium hirsutum* in response to the cotton bollworm *Helicoverpa armigera*. *Scientific Reports*, 5: 11867.

Hyten, D. L., Song, Q., Zhu, Y., Choi, I. Y., Nelson, R. L., Costa, J. M., Specht, J. E., Shoemaker, R. C. & Cregan, P. B. (2006). Impacts of genetic bottlenecks on soybean genome diversity. *Proceedings of the National Academy of Sciences*, 103(45):, 16666-16671.

Ishimoto, M., Sato, T., Chrispeels, M. J. & Kitamura, K. (1996). Bruchid resistance of transgenic azuki bean expressing seed α-amylase inhibitor of common bean. *Entomologia Experimentalis et Applicata*, 79(3): 309-315.

Kang, J. H., Wang, L., Giri, A. & Baldwin, I. T. (2006). Silencing threonine deaminase and JAR4 in *Nicotiana attenuata* impairs jasmonic acid–isoleucine–mediated defenses against *Manduca sexta*. *The Plant Cell*, 18(11): 3303-3320.

Kanrar, S., Venkateswari, J., Kirti, P. & Chopra, V. (2002). Transgenic Indian mustard (*Brassica juncea*) with resistance to the mustard aphid (*Lipaphis erysimi* Kalt.). *Plant Cell Reports*, 20(10): 976-981.

Kempema, L. A., Cui, X., Holzer, F. M. & Walling, L. L. (2007). Arabidopsis transcriptome changes in response to phloem-feeding silver leaf whitefly nymphs, similarities and distinctions in responses to aphids. *Plant Physiology*, 143(2): 849-865.

Khatodia, S., Bhatotia, K., Passricha, N., Khurana, S. M. P. & Tuteja, N. (2016). The CRISPR/Cas genome-editing tool: application in improvement of crops. *Frontiers in Plant Science*, 7: 506.

Kim, H. & Kim, J. S. (2014). A guide to genome engineering with programmable nucleases. *Nature Reviews Genetics*, 15(5): 321-334.

Kliebenstein, D. (2009). Quantitative genomics: analyzing intraspecific variation using global gene expression polymorphisms or eQTLs. *Annual Review of Plant Biology*, 60: 93-114.

Konno, K., Hirayama, C., Nakamura, M., Tateishi, K., Tamura, Y., Hattori, M. & Kohno, K. (2004). Papain protects papaya trees from herbivorous insects: role of cysteine proteases in latex. *The Plant Journal*, 37(3): 370-378.

Keurentjes, J. J., Koornneef, M. & Vreugdenhil, D. (2008). Quantitative genetics in the age of omics. *Current Opinion in Plant Biology*, 11(2): 123-128.

Kumari, R. & Kumar, S. (2015). Occurrence of molecularly diverse Bt Cry toxin-resistant mutations in insect pests of Bt+ corn and cotton crops and remedial approaches. *Indian Academy of Sciences,* 108(8): 1483-1490.

Lawrence, S. D. & Novak, N. G. (2006). Expression of poplar chitinase in tomato leads to inhibition of development in Colorado potato beetle. *Biotechnology Letters*, 28(8): 593-599.

Leplé, J. C., Bonadé-Bottino, M., Augustin, S., Pilate, G., Lê Tân, V. D., Delplanque, A., Cornu, D. & Jouanin, L. (1995). Toxicity to *Chrysomela tremulae* (Coleoptera: Chrysomelidae) of transgenic poplars expressing a cysteine proteinase inhibitor. *Molecular Breeding*, 1(4): 319-328.

Liu, X., Bai, J., Huang, L., Zhu, L., Liu, X., Weng, N., Reese, J. C., Harris, M., Stuart, J. J. & Chen, M. S. (2007). Gene expression of different wheat genotypes during attack by virulent and avirulent Hessian fly (*Mayetiola destructor*) larvae. *Journal of Chemical Ecology*, 33(12): 2171-2194.

Macedo, M. L. R., Freire, M. D. G. M., da Silva, M. B. R. & Coelho, L. C. B. B. (2007). Insecticidal action of *Bauhinia monandra* leaf lectin (BmoLL) against *Anagasta kuehniella* (Lepidoptera: Pyralidae), *Zabrotes subfasciatus* and *Callosobruchus maculatus* (Coleoptera: Bruchidae). *Comparative Biochemistry and Physiology Part A: Molecular & Integrative Physiology*, 146(4): 486-498.

Maheswaran, G., Pridmore, L., Franz, P. & Anderson, M. A. (2007). A proteinase inhibitor from *Nicotiana alata* inhibits the normal development of light-brown apple moth, *Epiphyas postvittana* in transgenic apple plants. *Plant Cell Reports*, 26(6): 773-782.

Majumder, P., Mondal, H. A. & Das, S. (2005). Insecticidal activity of *Arum maculatum* tuber lectin and its binding to the glycosylated insect gut receptors. *Journal of Agricultural and Food Chemistry*, 53(17): 6725-6729.

Mallikarjuna, N., Sharma, H. C. & Upadhyaya, H. D. (2007). Exploitation of wild relatives of pigeonpea and chickpea for resistance to *Helicoverpa armigera*. *Jounal of SAT Agricultural Research*, 3(1): 4.

Mao, Y. B., Cai, W. J., Wang, J. W., Hong, G. J., Tao, X. Y., Wang, L. J., Huang, Y. P. & Chen, X. Y. (2007). Silencing a cotton bollworm P450 monooxygenase gene by plant-mediated RNAi impairs larval tolerance of gossypol. *Nature Biotechnology*, 25(11): 1307-1313.

McManus, M. T., Burgess, E. P. J., Philip, B., Watson, L. M., Laing, W. A., Voisey, C. R. & White, D. W. (1999). Expression of the soybean (Kunitz) trypsin inhibitor in transgenic tobacco: Effects on larval development of *Spodoptera litura*. *Transgenic Research*, 8(5): 383-395.

Mooney, K. A., Halitschke, R., Kessler, A. & Agrawal, A. A. (2010). Evolutionary trade-offs in plants mediate the strength of trophic cascades. *Science*, 327(5973): 1642-1644.

Morton, R. L., Schroeder, H. E., Bateman, K. S., Chrispeels, M. J., Armstrong, E. & Higgins, T. J. (2000). Bean α-amylase inhibitor 1 in transgenic peas (*Pisum sativum*) provides complete protection from pea weevil (*Bruchus pisorum*) under field conditions. *Proceedings of the National Academy of Sciences*, 97(8): 3820-3825.

Oerke, E. C. (2006). Crop losses to pests. *The Journal of Agricultural Science*, 144(01): 31-43.

Pandey, G., Yadav, C. B., Sahu, P. P., Muthamilarasan, M. & Prasad, M. (2017). Salinity induced differential methylation patterns in contrasting cultivars of foxtail millet (*Setaria italica* L.). *Plant Cell Reports*, 36(5): 1-14.

Parde, V. D., Sharma, H. C. & Kachole, M. S. (2012). Protease inhibitors in wild relatives of pigeonpea against the cotton bollworm/legume pod borer, *Helicoverpa armigera. American Journal of Plant Sciences*, 3(5): 627.

Pautot, V., Holzer, F. M., Reisch, B. & Walling, L. L. (1993). Leucine aminopeptidase: an inducible component of the defense response in *Lycopersicon esculentum* (tomato). *Proceedings of the National Academy of Sciences*, 90(21): 9906-9910.

Pechan, T., Cohen, A., Williams, W. P. & Luthe, D. S. (2002). Insect feeding mobilizes a unique plant defense protease that disrupts the peritrophic matrix of caterpillars. *Proceedings of the National Academy of Sciences*, 99(20): 13319-13323.

Pechan, T., Ye, L., Chang, Y. M., Mitra, A., Lin, L., Davis, F. M., Williams, W. P. & Luthe, D. S. (2000). A unique 33-kD cysteine proteinase accumulates in response to larval feeding in maize genotypes resistant to fall armyworm and other Lepidoptera. *The Plant Cell*, 12(7): 1031-1040.

Ramesh, S., Nagadhara, D., Reddy, V. D. & Rao, K. V. (2004). Production of transgenic indica rice resistant to yellow stem borer and sap-sucking insects, using super-binary vectors of *Agrobacterium tumefaciens*. *Plant Science*, 166(4): 1077-1085.

Ramu, S. V., Rohini, S., Keshavareddy, G., Gowri Neelima, M., Shanmugam, N. B., Kumar, A. R. V., Sarangi, S. K., Ananda Kumar,

P. & Udayakumar, M. (2012). Expression of a synthetic cry1AcF gene in transgenic pigeon pea confers resistance to *Helicoverpa armigera. Journal of Applied Entomology*, 136(9): 675-687.

Rao, K. V., Rathore, K. S., Hodges, T. K., Fu, X., Stoger, E., Sudhakar, D., Williams, S., Christou, P., Bharathi, M., Bown, D. P., Powell, K. S., Spence, J., Gatehouse, A. M. R. & Gatehouse, J. A. (1998). Expression of snowdrop lectin (GNA) in transgenic rice plants confers resistance to rice brown planthopper. *The Plant Journal*, 15(4): 469-477.

Reymond, P., Bodenhausen, N., Van Poecke, R. M., Krishnamurthy, V., Dicke, M. & Farmer, E. E. (2004). A conserved transcript pattern in response to a specialist and a generalist herbivore. *The Plant Cell*, 16(11): 3132-3147.

Rhoades, D. F. & Cates, R. G. (1976). Toward a general theory of plant antiherbivore chemistry. *In*: Biochemical Interaction between Plants and Insects Springer, US. pp. 168-213.

Ryan, C. A. (2000). The systemin signaling pathway: differential activation of plant defensive genes. *Biochimica et Biophysica Acta*, 1477(1): 112-121.

Saha, P., Majumder, P., Dutta, I., Ray, T., Roy, S. C. & Das, S. (2006). Transgenic rice expressing *Allium sativum* leaf lectin with enhanced resistance against sap-sucking insect pests. *Planta*, 223(6): 1329.

Sarmah, B. K., Moore, A., Tate, W., Molvig, L., Morton, R. L., Rees, D. P. & Higgins, T. J. V. (2004). Transgenic chickpea seeds expressing high levels of a bean α-amylase inhibitor. *Molecular Breeding*, 14(1): 73-82.

Sharma, H. C., Sujana, G. & Rao, D. M. (2009). Morphological and chemical components of resistance to pod borer, *Helicoverpa armigera* in wild relatives of pigeonpea. *Arthropod-Plant Interactions*, 3(3): 151-161.

Stamp, N. (2003). Out of the quagmire of plant defense hypotheses. *The Quarterly Review of Biology*, 78(1): 23-55.

Steppuhn, A. & Baldwin, I. T. (2007). Resistance management in a native plant: nicotine prevents herbivores from compensating for plant protease inhibitors. *Ecology Letters*, 10(6): 499-511.

Stout, M. J., Riggio, M. R. & Yang, Y. (2009). Direct induced resistance in *Oryza sativa* to *Spodoptera frugiperda*. *Environmental Entomology*, 38(4): 1174-1181.

Sujana, G., Sharma, H. C. & Rao, D. M. (2008). Antixenosis and antibiosis components of resistance to pod borer *Helicoverpa armigera* in wild relatives of pigeonpea. *International Journal of Tropical Insect Science*, 28(04): 191-200.

Tscharntke, T., Thiessen, S., Dolch, R. & Boland, W. (2001). Herbivory, induced resistance, and interplant signal transfer in *Alnus glutinosa*. *Biochemical Systematics and Ecology*, 29(10): 1025-1047.

Turlings, T. C., Wäckers, F. L., Vet, L. E., Lewis, W. J. & Tumlinson, J. H. (1993). Learning of host-finding cues by hymenopterous parasitoids. *In*: Insect Learning Springer, US. pp. 51-78.

Vandenborre, G., Groten, K., Smagghe, G., Lannoo, N., Baldwin, I. T. & Van Damme, E. J. (2010). *Nicotiana tabacum* agglutinin is active against Lepidopteran pest insects. *Journal of Experimental Botany*, 61(4): 1003-1014.

Vandenborre, G., Miersch, O., Hause, B., Smagghe, G., Wasternack, C. & Van Damme, E. J. (2009). *Spodoptera littoralis*-induced lectin expression in tobacco. *Plant and Cell Physiology*, 50(6): 1142-1155.

Verhoeven, K. J., Jansen, J. J., Van Dijk, P. J. & Biere, A. (2010). Stress-induced DNA methylation changes and their heritability in asexual dandelions. *New Phytologist*, 185(4): 1108-1118.

War, A. R., Paulraj, M. G., Ahmad, T., Buhroo, A. A., Hussain, B., Ignacimuthu, S., & Sharma, H. C. (2012a). Mechanisms of plant defense against insect herbivores. *Plant Signaling & Behavior*, 7(10): 1306-1320.

War, A. R., Paulraj, M. G., War, M. Y. & Ignacimuthu, S. (2012b). Differential defensive response of groundnut germplasms to *Helicoverpa armigera* (Hubner) (Lepidoptera: Noctuidae). *Journal of Plant Interactions*, 7(1): 45-55.

White, C. & Eigenbrode, S. D. (2000). Effects of surface wax variation in *Pisum sativum* on herbivorous and entomophagous insects in the field. *Environmental Entomology*, 29(4): 773-780.

Yencho, G. C., Cohen, M. B. & Byrne, P. F. (2000). Applications of tagging and mapping insect resistance loci in plants. *Annual Review of Entomology*, 45(1): 393-422.

Zhao, L. Y., Chen, J. L., Cheng, D. F., Sun, J. R., Liu, Y. & Tian, Z. (2009). Biochemical and molecular characterizations of *Sitobion avenae*-induced wheat defense responses. *Crop Protection*, 28(5): 435-442.

Zhu, S. K., Salzman, R. A., Ahn, J. E. & Koiwa, H. (2004). Transcriptional regulation of sorghum defense determinants against a phloem-feeding aphid. *Plant Physiology*, 134(1): 420-431.

INDEX

A

B

C

D

E

F

G

H

I

L

M

N

O

S

T

U

V

W

Y

Z